Health and Safety in Small Industry

A PRACTICAL GUIDE FOR MANAGERS

Division of Consumer Health Education
Department of Environmental and Community Medicine
UMDNJ – Robert Wood Johnson Medical School

Library of Congress Cataloging-in-Publication Data

Health and safety in small industry.

 Bibliography: p.
 1. Industrial hygiene. 2. Industrial safety.
I. Robert Wood Johnson Medical School. Division of
Consumer Health Education.
RC967.H415   1989   613.6'2   88-26854
ISBN 0-87371-195-5

LEWIS PUBLISHERS, INC.
121 South Main Street, P.O. Drawer 519, Chelsea, Michigan 48118

PRINTED IN THE UNITED STATES OF AMERICA

New Jersey: Innovative Pioneer in Environmental Protection

New Jersey leads the way in numerous areas of environmental protection. As the fifth smallest state, and the most densely populated one, it has good reason for concern about its environment. Among the top seven producers of chemicals, pharmaceuticals, and petroleum distillates, it has had its full share of environmental problems. New Jersey tackles its problems, and continues with its pioneering efforts.

For example, New Jersey has established the first statewide mandatory recycling law in the United States. The state's "Spill Fund" tax for toxic cleanups was used as a model for the federal Superfund. New Jersey's Right to Know law for chemical workers and residents was the forerunner of recently established federal rules. Its aggressive testing for carcinogenic radon gas has led the nation; it includes testing of homes for radon gas, free of charge. In New Jersey, companies are required to rid sites of dangerous chemicals before they are allowed to sell the property, a move that has proved highly effective.

Other states have begun to look to New Jersey for leadership.

New Jersey's long-term approach to environmental protection includes, among many other innovations, this practical guide for managers in small industry.

This book is being made available worldwide, so that others may benefit from the innovations of New Jersey.

—The Publisher

Foreword

Assurance of the health and safety of employees is a primary responsibility of employers. Under the influence of economics, law, and, at times, conscience, major strides have been made to improve working conditions, prevent accidents, and reduce exposure to hazardous substances. The results, however, have been uneven. In the same town, you may find a workplace so clean "you could eat off the floor," while other workplaces are so contaminated you can hardly walk on the floor. More needs to be done to protect employees, and vigilance is required to assure that safety devices and programs currently in place continue to function optimally.

In many cases, employers have been reluctant to invest in safety and health measures because they can calculate the short-term costs but are uncertain about the long-term benefits of such efforts. In addition, insurance programs have not provided the strong incentive they might to encourage that the workplace be kept healthy and safe. To the chagrin of some and the delight of others, the federal government has used law, regulation, and enforcement to protect U.S. citizens at work. However, even this effort has been uneven from one year to the next and from one industry to the next.

In 1970, at the height of the Nixon Administration, the Occupational Safety and Health Act was passed by Congress and signed by the President. It provided for several different federal agencies to perform research, evaluate compensation insurance, and enforce regulations. Most of all, however, it established as a law of the land the responsibility of employers to provide "a safe and healthful workplace" for all employees.

This *Guide* is designed to help you understand your responsibilities under the law and to help you comply with the regulations which protect both you and your employees. It is a guide to statutes, regulations, and resources available to assist you. Various federal and state laws, such as the New Jersey Right to Know law and the federal hazard communication regulations that currently apply to small industries, are summarized. In addition, new areas of hazardous waste handling and toxic catastrophe prevention are examined. The responsibilities of relevant federal and state agencies are described and direct contact information is provided. The *Guide* also outlines a variety of medical, safety, worker education, and industrial hygiene resources that are available to employers, often at no cost, to assist in evaluating and correcting workplace hazards.

Despite newspaper accounts that the federal government has backed away from enforcing health and safety laws in the workplace, one must realize that the courts have not. Employers who have already invested in health and safety believe that failure of their competitors to make a comparable investment is unfair trade practice, and the courts have begun to take a dim view of such negligence. As more employers accept and act on their recognized responsibility to protect worker health and safety, there is increasing pressure on all to take appropriate measures.

In addition to the tangible information included, this *Guide* should provide encouragement to update health and safety programs and to provide the "safe and healthful workplace" as required by law and fairness.

Michael Gochfeld, MD, PhD
Clinical Professor and Chief
Division of Occupational Health
Department of Environmental and Community Medicine
Robert Wood Johnson Medical School
University of Medicine and Dentistry of New Jersey

Preface

Small companies and industries dominate the business landscape. In New Jersey, for example, small firms account for more than 97% of the state's business establishments. Varying in size from three to 100 employees, a typical small business might be a dry cleaning company, a tool and die shop, or a pharmaceutical firm. Regardless of size, however, the key asset of any firm is its work force.

In order to determine the needs of managers of small industries to provide a safe and healthy work environment, the Division of Consumer Health Education in the Department of Environmental and Community Medicine at the University of Medicine and Dentistry of New Jersey–Robert Wood Johnson Medical School conducted a survey of small industries in New Jersey. As anticipated, the surveyors found that small firms do not have the resources to employ safety and health experts such as physicians, nurses, industrial hygienists, or safety engineers; consequently, the development of a firm's safety and health program is the responsibility of the manager.

Health and Safety in Small Industry: A Practical Guide for Managers has been developed in response to the needs expressed by the respondents of the New Jersey survey, providing a desktop reference that addresses safety and health issues frequently confronted by managers of small firms. Although the *Guide* uses New Jersey as the model, the safety and health issues discussed are of national concern. The book has been prepared to offer guidance on the following topics:

- strategies to eliminate or control potentially hazardous situations
- laws that address occupational safety and health issues
- procedures to establish safety and health programs
- locating appropriate resources at minimum cost through voluntary nonprofit agencies, trade associations, employer groups, and government agencies

Because the needs for this reference were identified by managers of small companies and the content has been reviewed by distinguished colleagues, we are confident that *Health and Safety in Small Industry: A Practical Guide for Managers* will be a valuable resource for the development of health and safety programs by small firms.

Audrey R. Gotsch, DrPH
Associate Professor and Chief
Division of Consumer Health Education
Robert Wood Johnson Medical School
University of Medicine and Dentistry of New Jersey

Acknowledgments

A broad-based group of institutions supported the development of this *Guide* through the Foundation of the University of Medicine and Dentistry of New Jersey. For their interest and support of this reference, gratitude is extended to Exxon Corporation, Mobil, the Allied Foundation, Allied Signal, the Fund of New Jersey, American Hoechst Corporation, American Cyanamid, the Klipstein Foundation, the Dodge Foundation, the Winfield Baird Foundation, and the *Home News*.

Throughout the development of the *Guide,* the Division of Consumer Health Education has been guided by an Advisory Committee as well as a Task Force chaired by members of the Advisory Committee. For their assistance and encouragement, gratitude is extended to the following individuals:

Advisory Committee

Senator William Bradley
Honorary Chairman

Arthur Krosnick, MD, Chairman
Formerly, Editor, *New Jersey
   Medicine*

Ray Bramucci
Director of New Jersey Operations
   for Senator Bill Bradley

Edward W. Callahan
Vice President
Health, Safety, and Environmental
   Health
Allied Signal

Ronald Cohen, PhD
Health Officer
Middle Brook Regional Health
   Commission

Ronald J. Czajkowski, MA
Director of Communications
Center for Health Affairs
New Jersey Hospital Association

Richard T. Dewling, PhD
Formerly, Commissioner
New Jersey Department of
   Environmental Protection

Allan H. Doane
Vice President
Corporate Quality Assurance and
   Environmental Compliance
Warner-Lambert Company

Kolman J. Hahn, MS
Senior Staff Industrial Hygienist
Research and Environmental Health
   Division
Exxon Biomedical Sciences

Elissa Santoro, MD
Formerly, President, New Jersey
    Division
American Cancer Society

Robert Snyder, PhD
Director and Chairman
Department of Pharmacology and
    Toxicology
College of Pharmacy
Rutgers, The State University of
    New Jersey

Jeanne M. Stellman, PhD
Executive Director
Women's Occupational Health
    Resource Center
Associate Professor
School of Public Health
Columbia University

Michael Utidjian, MD
Corporate Medical Director
American Cyanamid Company

Task Force

A. Bernard Lindemann, Co-Chair
President
Miranol Chemical Company, Inc.

Anthony Mazzochi, Co-Chair
Formerly, Vice President
District 8 Council
Oil, Chemical, and Atomic Workers
    International Union, AFL-CIO
Institute for Labor Education and
    Research

Chee Keung Chan, PhD
Formerly, Director of Research and
    Development
Oakite Products

Mike McKowne
Formerly, Chairman, Occupational
    Health Committee and Health and
    Safety Representative for the
    UAW
American Lung Association of New
    Jersey

Edward Tinebra
Vice President of Technical
    Research and Development
J.L. Prescott Company

Authors

We particularly want to recognize the contributions that were made by the authors of this *Guide*. Primary manuscripts were provided by:

Cindy Rovins, MPH
Environmental Health Educator
Environmental and Occupational Health Information Program
Division of Consumer Health Education

Department of Environmental and Community Medicine
UMDNJ–Robert Wood Johnson Medical School

William Goldfarb, JD, PhD
Professor
Department of Environmental Resources
Cook College
Rutgers, The State University of New Jersey

In addition, appreciation is extended to Cindy Rovins, who served as the editor with Audrey R. Gotsch, DrPH, for the *Guide;* to Paul Landsbergis, EdD, Assistant Professor, Division of Consumer Health Education, Department of Environmental and Community Medicine, who wrote Chapter 2, Section 3, "New Jersey Right to Know"; to Constance Grzelka, BJ, Freelance Editorial Services, for her writing assistance; to Amy Duff, MHS, and Michele Demak, MPH, formerly with the Division of Consumer Health Education, for their contribution to the concept of the *Guide* and the implementation of the New Jersey survey; and to Carol Wimmer and Patricia Billman, Division of Consumer Health Education, for their clerical assistance.

Editorial Review Board

Gratitude is also extended to those who served as reviewers for the manuscript:

Edward D. Baretta, MS, CIH
Director, Safety and Industrial
   Hygiene
Johnson & Johnson

Donna M. Capizzi, CIH
University Industrial Hygienist
Department of Radiation and
   Environmental Health and Safety
Rutgers, The State University of
   New Jersey

Michael Gochfeld, MD, PhD
Clinical Professor and Chief,
   Division of Occupational Health

Department of Environmental and
   Community Medicine
UMDNJ–Robert Wood Johnson
   Medical School

Howard Kipen, MD, MPH
Assistant Professor
Division of Occupational Health
Department of Environmental and
   Community Medicine
UMDNJ–Robert Wood Johnson
   Medical School

Paul Landsbergis, EdD
Assistant Professor

Division of Consumer Health
Education
Department of Environmental and
Community Medicine
UMDNJ–Robert Wood Johnson
Medical School

Janice Marshall, RN, MSN
Acting Coordinator
Cancer Control and Risk Reduction
Program
New Jersey State Department of
Health

Molly J. McCauley, RN
Director, Total Life Concept
Manager, Health Promotion
AT&T

Lawrence Muzyka, CIH, CSP
Health and Safety Training
Coordinator
E.R. Squibb & Sons

Mitchel Rosen, MS
Course Coordinator
NJ/NY Hazardous Materials Worker
Training Center
Division of Consumer Health
Education

Department of Environmental and
Community Medicine
UMDNJ–Robert Wood Johnson
Medical School

Barbara Sergeant
Senior Environmental Specialist
New Jersey Department of
Environmental Protection
Bureau of Hazardous Substances
Information

James W. Stanley
Regional Administrator for U.S.
Department of Labor, OSHA,
Region II

Richard Willinger, JD
Manager, Right to Know Program
New Jersey State Department of
Health

Michael S. Zachowski
Formerly, Chief
Bureau of Emergency Response
New Jersey Department of
Environmental Protection

About the Environmental and Occupational Health Information Program

The Division of Consumer Health Education within the Department of Environmental and Community Medicine, Robert Wood Johnson Medical School (formerly Rutgers Medical School), University of Medicine and Dentistry of New Jersey engages in activities to promote educational opportunities for health care professionals and to encourage consumers to take a more active role in their own health care.

The Environmental and Occupational Health Information Program within the Division of Consumer Health Education was developed as a response to the need for public information regarding environmental and occupational health risks. This model program provides information and educational services to the general public, lay and professional employees, small industry, schools, and physicians.

Table of Contents

Publisher's Preface   iii
Foreword   v
Preface   vii
Acknowledgments   ix
About the Program   xiv

Chapter One: Overview of Occupational Safety and Health   1
1. Introduction   1
2. Striving for Safe Conditions in the Workplace   3
    Factors Affecting Safety   3
    Taking a Critical Look at Work Conditions   4
    Controlling Safety and Health Hazards in the Workplace   5

Chapter Two: Occupational Legislation   9
1. The Occupational Safety and Health Act   9
    OSHA Inspection Process   10
    Employee Rights Under OSHA   12
    Employee Responsibilities   13
    Employer Responsibilities   13
    Violations   14
    Citations/Penalties   15
    Appeals   15
    OSHA Standards   15
2. OSHA's Hazard Communication Standard   16
    History   17
    Implementation   17
3. New Jersey Right to Know   21
    State Right to Know Laws   21
    New Jersey's Worker and Community Right
       to Know Act   21
4. SARA Title III: Emergency Planning and
       Community Right-to-Know   34
5. Smoking in the Workplace   36
    Smoking in the Industrial Setting   36
    Legislation on Smoking in New Jersey   37

Chapter Three: Environmental Legislation 41

   1. Overview 41

   2. Federal Statutes 41

      The Clean Water Act 42

      The Clean Air Act 45

      The Toxic Substances Control Act 47

      The Resource Conservation and Recovery Act 48

      Section 311 of the Clean Water Act (Oil Spills) 51

      Comprehensive Environmental Response, Compensation and
        Liability Act 53

      SARA Title III: Emergency Planning and Community Right-
        to-Know Act of 1986 56

   3. New Jersey Statutes 57

      Toxic Catastrophe Prevention Act 57

      Spill Compensation and Control Act 58

      Environmental Cleanup Responsibility Act 59

**Chapter Four: Employee Safety and Health
and Industrial Hygiene 61**

   1. Employee Health 61

      Medical Surveillance 62

      Regulations Governing Employee Access to Monitoring and
        Medical Records 63

   2. Industrial Hygiene 64

      Physical Forms of Hazardous Substances 64

      Chemical Risk 65

      Routes of Exposure 65

      Effects of Hazardous Substances 66

      Recognizing, Measuring, and Evaluating Employee
        Exposure 66

      Industrial Hygiene Monitoring 68

   3. Personal Protective Equipment 72

      Head Protection 74

      Eye and Face Protection 75

      Protective Clothing 76

      Foot Protection 78

      Lifelines and Safety Belts 78

      Ear Protection 79

   4. Respiratory Protection 80

      Types of Respirators 82

      Selection and Fitting of Respirators 85

      Training Employees in Respirator Use 86

Maintenance of Respirators   87
Respirator Limitations   87
OSHA Minimal Acceptable Respirator Program   88
5. Employee Education and Training   89
6. Safety and Health Committees   92
Committee Structure   92
Role of the Safety and Health Committee   93
Guidelines   93
7. Promoting Employee Health   95
Implementing a Wellness Program   96
Illness Cost Audit: The Overall Picture   98
8. Sanitation   102
Design and Procedures   102

Chapter Five: Sources of Information and Assistance   105
1. Health and Safety Core Reference Library   105
2. OSHA Onsite Consultation for the Employer   108
Benefits   108
Procedure   109
Summary   111
OSHA/State Consultation Project Directory   112
3. OSHA Voluntary Protection Program   117
Program Description   118
Eligible Applicants   119
Program Options   119
Program Responsibilities   120
4. NIOSH Health Hazard Evaluation Program   121
5. Accredited Laboratories   122
6. Clinical Facilities for Evaluating Occupational Illness   123
7. Employee Assistance Programs   123
8. New Jersey's Workers' Compensation Law   124
9. Small Business Association and SCORE Offices   125
New Jersey SCORE Locations   126
10. Small Business Resources in New Jersey   127
New Jersey Small Business Development Center   127
New Jersey Office of Small Business Assistance   128
New Jersey Business Libraries   129

Appendix A: Resources for Chapter One   131
Occupational Health Resources   131
Selected Publications   131

National Nonprofit Organizations   131
Committees on Occupational Safety and Health (COSH)   134
New Jersey Agencies   136

Appendix B: Resources for Chapter Two   139
1. Occupational Safety and Health Resources   139
   How to Obtain OSHA Standards   139
   Selected Publications   140
   OSHA Offices   140
   NIOSH   148
   OSHA Hazard Communication Standard: Summary   149
2. New Jersey Right to Know Resources   150
   New Jersey Right to Know Law: Summary   150
   Obtaining Copies of the New Jersey Right to Know Act   152
   How to Obtain Hazardous Substance Fact Sheets   152
   Right to Know Governmental Information Resources   153
   County Lead Agencies   154
   Understanding and Using MSDSs   155
3. SARA Title III Resources   158
   National   158
   New Jersey   159
4. Smoking in the Workplace Resources   159
   National   159
   New Jersey   160

Appendix C: Resources for Chapter Three   163
1. Federal Environmental Legislation Resources   163
   General Publications   163
   General Information   164
   Clean Water Act   165
   Clean Air Act   165
   Toxic Substances Control Act   165
   Resource Conservation and Recovery Act   167
   CERCLA (Superfund)   168
   SARA Title III   168
2. New Jersey Environmental Legislation Resources   169
   Toxic Catastrophe Prevention Act   169
   Spill Compensation and Control Act   170
   Environmental Cleanup Responsibility Act   170
   DEP Hazardous Waste Advisement Program   170

3. U.S. EPA Regional Offices   170
    EPA Region II Directory   172
4. New Jersey DEP Directory   173

Appendix D: Resources for Chapter Four   175
    1. Employee Education and Training Resources   175
        Suggested Readings in Industrial Safety and Health
            Training   175
        Companies Producing Materials for Employee Education and
            Training   176
        Federal Agencies   177
        Organizations   180
    2. Employee Health Promotion Resources   188
        Selected Publications   188
        Nonprofit Organizations: National and New Jersey   189

Appendix E: General Information   195
    1. Standard U.S. Federal Regions   195
    2. Steps for Responding to a Chemical Emergency   196
        General Guidelines   196
        Suspected Radioactive Material   196
        New Jersey Procedures   197
    3. Glossary of Occupational Health Terms   198

Index   201

CHAPTER 1

Overview of Occupational Safety and Health

1. INTRODUCTION

Creating a safe work environment is a matter of prevention, which is the primary way to avoid illness, accidents, and deaths in the workplace. Injuries and illnesses cost business a great deal in both direct and indirect costs, such as workers' compensation, medical insurance, interrupted production, wages paid for lost time, and time spent processing forms. Preventing injuries and illnesses is a good business investment.

Public concern for safety began with the start of the Industrial Revolution in the United States. As industry grew, so did the number of machines, toxic materials, and hazardous working conditions. The efforts of labor organizations and reform groups led to growing public awareness that has produced many safety and health improvements for the work force over the years. In the early 1900s, the U.S. Department of Labor was established, and the federal government created a division of occupational health within the U.S. Public Health Service. Workers' compensation laws were adopted by the states to provide some economic protection to injured workers and to give employers an incentive to maintain safe working conditions.

The Occupational Safety and Health Act (OSHAct) of 1970 established a comprehensive occupational safety and health law "to assure so far as possible for every working man and woman in the nation safe and healthful working conditions and to preserve our human resources."

In passing this law, Congress also created a regulatory agency, the Occupational Safety and Health Administration (OSHA), that was charged with monitoring the five million workplaces in the nation. OSHA is responsible for establishing and enforcing minimum rules and standards that regulate the health and safety of most of the nation's 97 million workers. It serves as the underlying support structure that enables workers to take an active role in correcting health and safety hazards.

The OSHA team of inspectors is small and visits fewer than 2% of all the nation's workplaces each year. Employees have the right to request an inspection by OSHA when they have a complaint about unsafe and unhealthy working conditions. A call for an OSHA probe, however, should usually be considered as a last resort when management has not responded to or corrected employee complaints about hazards.

Many workplace tragedies have claimed lives and many health hazards such as dust, fumes, and chemicals have silently eroded the health of countless American workers. The exact number of workers affected is not known. A recent survey of death certificates from all 50 states and the District of Columbia by the National Institute for Occupational Safety and Health (NIOSH) showed that at least 32,342 Americans died at work from 1980 to 1984.

The statistics don't enumerate the number of workers who will suffer from the agony of asbestosis, black lung disease, cancer, and other diseases. Because there is often a long period of latency following initial exposure to a carcinogen (cancer-causing substance), a worker may show no symptoms of disease until 20 or 30 years later. By then, the previous exposure has made him a victim of premature death or disability.

With careful attention to industrial process and design, work practices, and education, most toxic chemicals can be handled safely and exposure to most hazards can be minimized. However, exposure to toxic substances in the work setting can damage the human body in many ways. Some exposures are fatal, whereas others lead only to temporary changes in a particular organ. Some exposures may cause cancer in a worker years after daily handling of a hazardous substance, while others may alter the normal development of a fetus in a pregnant woman.

Technology has yielded many advances for business and industry, but it also has produced an array of potential hazards in the workplace. The list of new chemicals and new processes is constantly growing; it is estimated that 100,000 to 500,000 different materials are used by industry today. Because all workers have a right to know which chemicals and materials they use in their jobs and how these substances can affect their health, businesses are required to have a list of all chemicals and substances that are used on the premises, according to the 1985 OSHA Hazard Communication Standard. Managers are responsible for seeing that each container is clearly labeled and that appropriate training is provided for employees regarding the safe handling of industrial materials.

Not all of the problems encountered at work are chemical. Other sources of stress may affect health and can be minimized or avoided. These stressors include physical and infectious agents, ergonomics, work on video display terminals, psychosocial factors, fear of AIDS, etc.

2. STRIVING FOR SAFE CONDITIONS IN THE WORKPLACE

The work environment potentially has a tremendous effect, both good and bad, on the health of workers. Injury and disease in the workplace can be eliminated through prevention. Because you are most familiar with your own business, you are probably your own best safety analyst. This book will provide you with more specific skills to examine your workplace. The following general guidelines are a good first step to finding and correcting hazards in your business.

Factors Affecting Safety

Make an in-house safety inspection and look at how your employees approach their work. Look at the condition of your equipment and the environment you provide. Consider how changes in processes might alter hazards. As you review the work setting, the following factors should be considered:

1. *Employees:* Are the workers properly trained to do their jobs safely? Does training of new employees begin the first day on the job? Are employees only concerned about production? Do they combine good safety practices with their production skills? Do your managers understand the importance of health and safety?
2. *Equipment:* Is your equipment maintained properly? Is anything defective? Is it right for the job? Does it meet safety standards? Is it designed so that workers cannot remove or bypass any safety guards? Are switches and wires in good condition?
3. *Environment:* What are the conditions of the plant? Is there a lot of dust, fumes, and/or noise? Does the facility meet fire, electrical, and building codes and standards? Are there smoke detectors? Are work positions comfortable? Are floors, stairs, railings, and aisles clean and safe? Is first aid equipment within easy reach of the work areas and is someone trained to use it?
4. *Yourself:* Are you familiar with the basic fire, electrical, and building codes and standards that minimize safety hazards? Are you familiar with new state and federal reporting requirements? Do you know where to go for information, advice, and technical assistance?

Employees should be involved in safety inspections. Their participation and suggestions can lead to better hazard control and accident prevention. One of the common approaches to preventing accidents and illnesses at work is the establishment of management-worker joint health and safety committees to regularly review and plan for health and safety in the workplace.

Taking a Critical Look at Work Conditions

Several factors influence the work environment and it is important to consider how each factor may affect your employees.

Consider the physical environment. Is it too hot or too cold? Is it damp? What is the level of noise and vibrations during a shift? Is the lighting adequate? Are stairways in good condition and protected with safety railings? Do aisles contain clutter and debris? Are fire exits well marked? Are fire extinguishers within easy reach? Constant use of equipment and facilities causes wear and tear on machines and the building itself. Substances emitted from machines or materials can create a buildup of dirt on light fixtures, on windows, and in ventilation systems. As a manager, you must be constantly vigilant about the upkeep of equipment and the work setting; in addition to being economically beneficial in many cases, it is your responsibility to provide a safe workplace.

The use of chemicals may be a big part of your business. If chemicals are used, you are required to keep a list of all substances for future reference. It is important to routinely take samples of the air to check for the presence of airborne contaminants. Airborne dust, vapors, and gases that are inhaled can enter the respiratory tract. Solid or liquid materials that touch the skin can be absorbed into the eyes or body. New chemical products are constantly being used to update and modify all kinds of work processes. When you consider using a new material, you must carefully evaluate the substance to determine how it should be used so that it will not have a toxic effect on the employees who use it. If an employee must work overtime, will the increased hours mean an increased exposure to a toxic substance? Do you take air samples whenever you change the chemicals used, change/add new equipment, or change procedures?

Workers can be exposed to biological contaminants, such as bacteria, viruses, and fungi, in certain occupations. For example, health care and laboratory workers come into contact with blood-contaminated needles and other objects that can produce infection if workers do not exercise care. Employees must be educated in the safe use and disposal of these objects to reduce the chance of infection. Farm workers, animal handlers, and slaughterhouse workers risk infection from viruses, bacteria, and parasites. To prevent infection, they may need to wear protective clothing, shoes, and in certain instances, respirators. Other examples include excavators and workers operating bulldozers, who risk fungal infection from inhaling dust that contains spores. If fungal contamination is present, a fungicide can be sprayed on the soil to help prevent disease. Florists and nursery workers also can contract fungal infections if they accidentally stick themselves with contaminated sphagnum moss. Again, infection can be prevented by spraying the moss with a fungicidal preparation. In general, health education and an emphasis on good hygiene can help control risk from exposure.

Stress created by a demanding job can take its psychological toll on a worker. Stress has been blamed for many physical ailments, including ulcers, recurring headaches, accidents, and alcohol and drug abuse. It also can have a negative influence on social relationships at work and at home. When a worker has a large workload and inadequate supplies and resources to do the job well, and is required to put in long hours, he or she may be under stress. Managers can help prevent work-related stress by matching employees with jobs that suit their skills, giving positive reinforcement, getting employees' input, delegating authority, and encouraging a social support system among employees and their families.

Controlling Safety and Health Hazards in the Workplace

Monitoring

Employers must strive constantly to protect their employees from any exposure to unsafe equipment and conditions. Towards this end, two kinds of monitoring are commonly performed: environmental monitoring and medical surveillance.

Environmental monitoring. If it is suspected that exposure has occurred, air sampling tests can be performed to analyze the air in the work area. Air samples should be taken by trained personnel or industrial hygiene specialists. Special attention should be given to monitoring the air concentrations when your manufacturing process is producing its highest level of emissions. Some measurements can be made by using direct-reading instruments that measure gases and vapors in the air. Other sampling methods use specially treated paper or badges that react when they come into contact with certain substances. Still other methods may require collecting an air sample in a tube or on a filter, with analysis performed by an analytical laboratory. You must be certain that the air sampling and analysis are specific for the substances to be monitored.

Medical surveillance. Medical examination of employees, including tests of blood and urine, can help determine whether a worker has been exposed to a hazardous substance. It is important that the examinations and tests performed are appropriate for the expected exposure. Many so-called occupational medical services support themselves by excessive and inappropriate laboratory tests.

Hazard Minimization

If tests show that a health hazard is present, several strategies can be employed to control the condition. But before using any of these methods, you must be sure that it is the best way to control the problem and that the solution doesn't create a problem elsewhere, such as the contamination of air, soil, or water near your plant. The most effective ways of controlling health hazards are described in the following list. (Many of these focus on chemical hazards and reducing chemical exposure. More technical information on allowable exposure limits is provided in Chapter 4, Section 2.)

1. *Substitution* can solve the problem if a less hazardous material or process is used in place of a more hazardous one (for example, replacing organic solvents with detergent and water, or replacing a carcinogenic solvent with a noncarcinogen). You also can substitute the process used for doing the job. For instance, spray painting can be substituted by brush painting so as to reduce the amount of airborne contaminants from toxic paints. The great reductions in industrial use of carbon tetrachloride, benzene, and asbestos are examples of successful substitution.

2. *Ventilation* can remove or control the buildup of toxic materials and send fresh air into the work area. If toxic emissions are low and far enough from the breathing zone of workers, then a general ventilation system can be used. A local exhaust ventilation system is necessary when highly toxic vapors, fumes, or dusts are released directly within the worker's breathing zone (for example, when filling drums or barrels). A local system uses a hood, ducts, fan, and air purifier to capture a contaminant at its source and draw the contaminated air away from the worker. Proper maintenance is imperative if a ventilation system is to work effectively. Local ventilation systems are usually inexpensive and can be immediately cost-effective.

3. *Isolation or enclosure* of the process limits worker contact with hazardous materials or operations. This can mean moving the hazardous job to a more isolated location within the plant where fewer people will be exposed. Enclosing the process can eliminate or greatly reduce the escape of dust or vapors into the work area. The worker also can be protected from noise, heat, and fumes by erecting a barrier or by enclosure in a specially designed booth.

4. *Housekeeping* and equipment maintenance duties must be performed on a strict schedule. Containers used for storing solvents and other toxic materials should be checked for leaks. Airtight metal containers should be used to dispose of rags containing solvents. These containers should be disposed of properly on a daily basis. Aisles should be free of clutter. Dust (such as asbestos or lead dust) should be controlled before it presents a problem. (For more infor-

mation, see entries above on substitution and ventilation.) A clean, neat, and orderly workplace has an effect on health as well as morale.

5. *Personal protective equipment* can help control contact with hazardous materials, but is not a first-line defense for workers. Such equipment is meant to be used as a last-resort protection in a temporary situation, such as accident prevention or cleanup of a toxic spill. (For a more detailed description, see Chapter 4, Section 3.)

6. *Mechanizing* the process can help workers in jobs where lifting large bundles and bins is required. By installing a mechanical lifting device, back and shoulder injuries could be greatly reduced.

Educational programs that emphasize awareness are an essential facet of controlling workplace hazards. The establishment of a joint management-worker safety committee is an important step. This is an excellent way of recognizing hazards and preventing costly accidents. (See "Safety and Health Committees," Chapter 4, Section 6.)

Most small businesses that use machines already have a safety plan in place. While each company should have its own safety and emergency response procedures, there are some elements that should be common to all safety policies:

- Strictly enforce safety rules that are considered a part of the job, such as the wearing of safety shoes or gloves. Encourage safe practices among employees. This goes hand-in-hand with safe design of equipment and processes.
- Explain fully and clearly the materials and processes involved in the job and the potential hazards in handling them.
- Train employees to properly handle any hazardous substances.
- Require that all accidents be reported so that the cause can be corrected and recurrence prevented.
- Promote concern and commitment on the part of management and workers to operate a facility that meets the highest safety standards.

As a small business operator, you have the advantage of close contact with your employees and a working knowledge of the problems in your company. You can protect your human assets by protecting their health and safety. In addition, as your employees see that you are trying to improve their work environment, they'll be encouraged to work with you to prevent accidents and injury at work.

References

"Traumatic Occupational Fatalities—United States, 1980–1984," Morbidity and Mortality Weekly Report (Washington, DC: U.S. Department of Health and Human Services/Public Health Service, Centers for Disease Control, July 24, 1987).

"What Every UAW Representative Should Know About Health and Safety" (Detroit, MI: International Union, United Auto Workers, UAW Health & Safety Department, July 1979).

Levy, B. S., and D. Wegman, eds. *Occupational Health: Recognizing and Preventing Work-Related Disease* (Boston: Little, Brown and Company, 1983).

Protecting Workers' Lives: A Safety and Health Guide for Unions (Chicago: National Safety Council, 1983).

"OSHA Handbook for Small Businesses" (Washington, DC: U.S. Department of Labor, Occupational Safety and Health Administration, revised 1979).

The Industrial Environment—Its Evaluation and Control (Washington, DC: U.S. Department of Health and Human Services/Public Health Service, National Institute for Occupational Safety and Health, 1973).

CHAPTER 2

Occupational Legislation

1. THE OCCUPATIONAL SAFETY AND HEALTH ACT

This information was adapted with permission from *Protecting Worker's Lives: a Safety and Health Guide for Unions,* National Safety Council, 1983.

Faced with rapidly developing technology and a work force subject to occupationally caused illness and death, the U.S. Congress responded by enacting a protective piece of legislation, the Occupational Safety and Health Act of 1970. The OSHAct requires that "each employer shall furnish to each of his or her employees a workplace which is free from recognized hazards that are causing or are likely to cause death or serious physical harm to his or her employees and shall comply with occupational safety and standards promulgated under this Act." This is widely interpreted as meaning the employer shall provide a "safe and healthful workplace."

From this legislation evolved three new government agencies:

1. OSHA, in the Department of Labor, created to set and enforce safety and health standards in workplaces
2. NIOSH, in the Department of Health and Human Services, a research organization
3. the Occupational Safety and Health Review Commission (OSHRC), which acts as an appeals court for enforcing OSHA regulations

The OSHAct prescribes minimum safety and health standards for most industries. It authorizes inspections, citations, and mandatory and civil penalties to enforce its standards.

Under the OSHAct, states may establish and administer their own occupational safety and health programs and agencies. OSHA funds up to 50% of each program's operating costs. State programs must provide protection to workers at least as effective as the federal program. In order to obtain federal approval for a state program, the state must also include coverage for public employees in the program. There are currently 24 state programs.

Most American workers are covered under the federal OSHAct, with the exception of federal, state, and local government employees, self-employed people, and workers in states with approved OSHA programs. Workers also exempted from OSHA are those covered by other laws and agencies, such as the Nuclear Regulatory Commission, the Mine Safety and Health Act, and the Department of Transportation.

OSHA sets a variety of standards to prevent accidents and hazardous exposures. A standard is a written specification to protect employees. It may describe a safeguard or establish an exposure limit.

The OSHAct provides for emergency temporary standards, advisory committees, data collection, training and education, research, and assistance to small businesses, and is empowered to shut down workplace operations that pose imminent danger to safety and health.

By law, the Occupational Safety and Health Administration publishes standards, inspects workplaces to see if such standards are being met, issues citations, and imposes penalties on employers if they are in violation of standards. Fines can range from $1000 to $10,000 per violation, depending on the type of violation. In reality, most fines are not imposed at the highest level. Employers can apply for variances from standards if they can prove that equal protection is provided in another way. Employers who have received citations or fines have the right to appeal to the Occupational Safety and Health Review Commission. Employers also have the right to appeal to the U.S. Federal Court.

OSHA Inspection Process

OSHA is authorized to conduct workplace inspections, but generally inspects less than 2% of all workplaces per year. *Although OSHA covers all workplaces, it doesn't perform generally scheduled inspections of workplaces with an average of 10 or fewer employees during the year.* The following is a general outline of the inspection process. For more information regarding inspections, contact your local OSHA office. (See resources in Appendix B.)

Schedule of Inspections

OSHA has set priorities for workplace inspections. In order of priority, workplaces are inspected as follows:

1. workplace situations that present an imminent danger of causing death or serious injury

2. after catastrophes and fatal accidents (All employers must report to OSHA, within 48 hours of occurrence, all fatal accidents and all accidents causing five or more employees to be hospitalized.)
3. after complaints by employees or their representative(s) (union, etc.)
4. as a scheduled, regular inspection of a workplace in a high-hazard industry
5. as a reinspection

In general, OSHA will only inspect workplaces after receiving a *formal complaint*. Formal complaints must be *written*, listing alleged violations and giving examples of such alleged violations. Formal complaints must also be signed by an employee currently employed, his/her representative (union, attorney, family member), or an elected official. OSHA may respond to telephone complaints if it is believed that the facts given pose an imminent danger and warrant an immediate inspection of the worksite. OSHA is required to respond to formal complaints within 24 hours in cases of imminent danger, within three days if the complaint is deemed serious, and within 20 days for other complaints. *Nonformal complaints* that do not meet the above criteria will not evoke an inspection. OSHA will send a letter to an employer, briefly describe the suspected violation, and will request compliance within a certain time. Complainants will also receive a copy of the letter and will be informed of OSHA's actions; they will be requested to notify OSHA should the hazard not be corrected or removed. If a complainant then informs OSHA that the hazard still exists, OSHA will send an inspector.

Inspection Requests

By law, employees have the right to request an inspection. All employees who have signed an OSHA complaint form (OSHA-7), or have signed a letter sent to OSHA, have the right to speak to the OSHA inspector. OSHA inspections consist of three parts: an opening conference, a walkaround inspection of the workplace, and a closing conference.

Opening conference. The opening conference is the first step of the inspection. Depending on the size of the workplace, inspections may take several hours or several weeks. In general, the OSHA compliance officer meets with a company representative and an employee representative to discuss the inspection procedure. The officer will explain the purpose of the visit, the scope of the inspection, and OSHA standards that apply. Any previous inspections will also be discussed. An employer may request a separate conference, and OSHA will provide written summaries of each conference to each party, upon request. If an officer believes that an immediate inspection is warranted, the officer may shorten the opening conference and inspect immediately. Employees have the right to request immediate inspection.

Walkaround inspection. This follows the opening conference. The officer will check the workplace for hazards listed in the complaint, accompanied by representatives from management and labor. The entire worksite may be inspected if it is a regular, scheduled inspection.

Depending on whether it is a safety or health inspection, the officer will inspect for certain hazards and may use various instruments to measure or monitor health hazards. During the opening conference or the inspection, the officer will check the employer's injury and illness records and any documentation of safety training. If there is no formal complaint, and the employer can document below-average illness and injury rates, the safety officer may leave the worksite without actual inspection. However, an industrial hygiene officer will always conduct an inspection regardless of the lost-work-day injury rate.

Generally, the officer has the right to visit any part of the workplace, even if a complaint cites a specific location, injury, or illness. Lack of a search warrant can limit an inspector's access to the worksite, but if obtained, the officer has the right to total access.

Closing conference. A closing conference is held after the inspection, with the OSHA officer, employer, and employee representative. Separate meetings may be requested. The officer discusses what has been found and indicates violations for which citations may be proposed or recommended. At this time, the employer is told of appeal rights. The officer does not indicate any proposed penalties—only the area OSHA director has that authority. The officer will also discuss time and methods needed to correct any violations that may be found. Employers have the right to request a longer abatement period to correct such violations. This request may be questioned by the officer and employee representative. Employee representatives have the right to request copies of any citations and results of any inspections, testing, and monitoring. They may also request attendance at any future meetings between OSHA and the employer.

Employee Rights Under OSHA

The rights of employees are specifically addressed in the OSHAct. They include the following:

- to have safety and health standards established (promulgated) and enforced by law
- to file a complaint with OSHA and to request inspection
- to remain anonymous after filing a complaint
- to have any employee representative (union) to accompany an inspector during inspection and to point out suspected hazards

- to inform the OSHA inspector (verbally or in writing) of suspected hazards
- for employees who have filed complaints, to receive a copy of any citations given and the time allotted for compliance
- to be informed if no violation or hazard exists, or if no citations have been issued. Employees have the right to an informal meeting with OSHA to discuss why no citations have been issued.
- to file a "Notice of Contest" with OSHA should employees disagree with the amount of abatement time granted to the employer to correct violations
- to attend any OSHA Review Commission meeting during employer appeal and to appeal Review Commission decisions to the U.S. Court of Appeals
- to be told about hazardous conditions that exceed permissible standards published after 1971
- to obtain a copy of the OSHA workplace file, copies of injury and illness records, and results of any testing for noise, dusts, and fumes
- not to be discharged or discriminated against for exercising any of the above rights under OSHA

Employee Responsibilities

Employees also have responsibilities under the OSHAct. They must comply with all OSHA regulations and standards, and must not remove, displace, or interfere with any safety and health safeguards in the workplace.

Employer Responsibilities

Employers must also comply with OSHA standards and regulations as follows:

- Follow all OSHA standards.
- Keep records of all injuries and occupational illnesses, make them available to employees upon request, and post a summary of these records annually, during the month of February.
- Permit an employee or representative to accompany an OSHA inspector during inspection, testing, or monitoring of the worksite.
- All employee rights under OSHA must be posted in the workplace.
- All employees must be warned of potential hazards via posters, labels, or color codes.
- Provide and pay for all testing (air monitoring, medical testing, etc.).
- Post in a prominent place, near the violation, a copy of any OSHA citations.
- Provide employees and/or their representatives with copies of medical and chemical exposure records.

OSHA 200 Log of Injuries and Illnesses

OSHA requires private employers to maintain the "200 Log," a record of work-related illnesses and injuries, and to make the records available to employees, former employees, or employee representatives upon request. Each February the summary page of the previous year's 200 Log must be posted in the workplace for employees to read.

The 200 Log has separate sections for injuries and illnesses. For each employee hurt on the job who has either lost time from work or received medical treatment, the following information must be recorded:

- case or file number
- date of injury or onset of illness
- employee's name
- occupation
- department
- description of injury or illness
- number of days away from work
- number of days of restricted work activity
- date of death, for fatalities

The information contained in the 200 Log is an important resource for analyzing trends or patterns in the types of injuries or illnesses occurring and the occupations and departments in which they are occurring. Accurate, up-to-date records are necessary if they are to be useful to employers and employees. In addition to the 200 Log, employers must maintain a "Supplementary Record of Injury or Illness," giving more details on each recorded case.

Violations

During an inspection, OSHA may find two kinds of violations. The first is a violation of any OSHA standard. The second is known as a "general duty clause" violation and is used to cite hazards not covered by a specific OSHA standard. Section 5(a)(1) of the OSHAct, the "general duty clause," requires that employers keep workplaces "free from recognized hazards that are causing or are likely to cause death or serious physical harm." For penalty purposes, violations are classified as "serious," "other than serious," "imminent danger," "willful," and "repeat."

Willful violations are rarely issued because willfulness is difficult to prove.

De minimis violations are those that do not pose any direct or immediate effect on safety and health; thus, no citations or penalties are issued for this violation.

Violations that *are not likely* to cause death or serious physical harm if an accident occurs are called *other than serious*. OSHA can propose a penalty of up to $1000 per violation. This penalty can be reduced, depending on certain conditions.

Violations that *are likely* to cause death or serious physical harm if an accident occurs and about which the company knew, or should have known, are called *serious*. OSHA must propose a penalty of up to $1000, which may also be reduced.

Willful or *repeat* violations of OSHA standards may be penalized by up to $10,000 per violation. Should a willful violation cause death, the maximum penalty is $10,000, imprisonment up to six months, or both. A second violation of this type doubles the penalty.

Should an employer not correct a violation within the amount of time allotted for such corrections, OSHA can issue Form 2-B, "Notification of Failure to Correct Alleged Violation." Additional penalties are calculated by the amount of the original penalty, multiplied by the number of days the violation has continued, up to a usual maximum of 10 days.

Citations/Penalties

Citations and penalties are determined by the OSHA area director after receipt of the report from the compliance officer. Notices of citations are usually mailed to an employer by certified mail, with one copy sent to the union. A citation informs the employer and employees of violations of regulations and standards, and states a time period to correct such violations. Employers must post a copy of each violation near the location of the violation. Posting must be for three days, or until the violation is corrected, whichever is longer. Where OSHA allows an employer three months to correct a violation, a union representative or safety and health committee member may request a copy of the employer's abatement plan and a copy of the three-month progress report required by OSHA.

Appeals

Employers may appeal any citation, penalty, and/or parts thereof. Employers must send a "Notice of Contest" to the local OSHA office. A union has the right to become part of this appeal by requesting "party status." Under party status, the union has a right to participate in any hearing, review all documents provided by the employer, cross-examine employer's witnesses, and present its own witnesses.

OSHA Standards

Standards, both advisory and mandatory, are those guidelines set up by OSHA that outline safe workplace design and performance in four major categories: General Industry, Maritime, Construction, and Agriculture. Standards are produced by organizations such as the American National Standards Institute (ANSI), the National Fire Protection Association (NFPA), and the American Conference of Governmental Industrial Hygienists (ACGIH).

Horizontal standards are those that apply to all workplaces, and are found in Title 29 ("Labor") of the *Code of Federal Regulations* (CFR), beginning with reference numbers 1910. (See resources in Appendix B.)

Vertical standards apply to a particular industry or trade such as shipbuilding, longshoring, construction, etc. If there is a conflict as to which standards apply, the vertical or special standards should be applied first. Standards are applied according to the type of work performed.

Health standards make up 34% of General Industry Standards, and regulate chemical, noise, and radiation exposures; requirements for ventilation; respirators; and personal protective equipment. Health standards also regulate recordkeeping and access to exposure and medical records.

Permanent standards are those adopted by OSHA. Such standards are established by advisory committees, usually the National Advisory Committee on Safety and Health and the Federal Advisory Committee on Safety and Health. Interested parties have the right to comment on permanent standards that will be promulgated.

Emergency temporary standards can be imposed by OSHA if it is believed that a substance or operation poses imminent danger and no current standard exists. These standards take effect immediately.

Voluntary standards are not enforced by law, and are usually promoted by manufacturers and users concerned about safety and health.

For information on obtaining copies of standards and occupational health resources, see Appendix B.

2. OSHA'S HAZARD COMMUNICATION STANDARD

This section on OSHA's Hazard Communication Standard was adapted with permission from "Highlights of OSHA's Hazard Communication Standard," OSHA, November 22, 1983, and OSHA regulation 29 CFR 1910.1200.

On November 25, 1983, OSHA published its Hazard Communication Standard (29 CFR 1910.1200). The standard went into effect on November 25, 1985. It was amended on August 24, 1987, to include all nonmanufacturers in the private sector, in addition to the manufacturers previously covered.

Under this regulation, chemical manufacturers and importers must evaluate or assess the hazards of chemicals which they produce or import, and, along with distributors, send that information to other manufacturing and nonmanufacturing employers through use of container labeling and Material Safety Data Sheets (MSDSs). Also, all manufacturing and nonmanufacturing employers must inform their employees about the hazardous chemicals to which they are exposed by means of a hazard communication program, labels, MSDSs, and training.

History

Since the passage of the OSHAct in 1970, comprehensive health standards covering 23 toxic substances have been adopted by OSHA including standards for asbestos, arsenic, ethylene oxide, benzene, lead, vinyl chloride, and cotton dust. The benzene standard was overturned by the Supreme Court, and was rewritten by OSHA (29 CFR 1910.1028, "Benzene," effective date December 10, 1987). All of these standards contain labeling provisions, since Section 6(b)(7) of the OSHAct specifically calls for labeling as an appropriate way to alert employees to hazards in the workplace.

In order to provide employees with greater knowledge about the health effects of toxic chemicals not covered by these comprehensive standards, OSHA enacted, on May 23, 1980, a regulation permitting workers to obtain copies of existing employer-maintained medical records and records of exposure to toxic substances (29 CFR 1910.20). This regulation covers records applying to all chemicals (over 70,000) contained in the Registry of Toxic Effects of Chemical Substances (RTECS) of NIOSH.

In addition, on January 13, 1981, OSHA proposed a standard requiring the labeling of hazardous chemicals used in manufacturing. OSHA pointed to NIOSH's 1972 National Occupational Hazard Survey, which indicated that one in every four workers (about 25 million) was exposed to hazardous substances. Trade name products accounted for 70% of all exposures, and for 90% of the products no information was available in the workplace to employers or employees identifying the chemicals in the products. Initial results of the second National Survey, begun in 1981, showed that 76% of all exposures were to trade name products. This proposal was withdrawn on February 12, 1981.

Implementation

Industries Covered

Employers in the manufacturing sector, nonmanufacturers, and importers and distributors of chemicals are covered. They have to have a written hazard communication program, label containers, keep Material Safety Data Sheets, and train employees. Chemical laboratories, however, are subject to reduced requirements, as follows:

- Keep labels on incoming hazardous chemical containers.
- Keep MSDSs and make them available to employees.
- Train employees as required by the Standard.

The Standard covers employees "who may be exposed to hazardous chemicals under normal operating conditions or foreseeable emergencies, including, but not limited to, production workers, line supervisors, and repair or maintenance personnel."

Chemicals Covered

Any chemical which is a physical hazard (combustible, explosive, flammable, reactive, an oxidizer, compressed gas, or organic peroxide) or a health hazard (irritant, corrosive, allergic sensitizer, cancer-causing, toxic, reproductive hazard, or any chemical that causes short-term or long-term damage to the body) is covered. The standard establishes a "floor" of approximately 2300 substances automatically covered (those regulated by OSHA, or listed by the American Conference of Governmental Hygienists, National Toxicology Program, or the International Agency for Research on Cancer). The standard covers all substances which may pose a hazard to human health. Beyond the 2300, the manufacturer or importer must evaluate all other chemicals to determine if they are hazardous, using certain criteria listed in the Standard.

Not covered by the Standard are wood; tobacco; potentially hazardous substances such as drugs, food, and cosmetics brought into the workplace for the personal use of employees; or hazardous wastes regulated by the U.S. Environmental Protection Agency (EPA).

Written Program

All manufacturing and nonmanufacturing employers must develop and carry out a written hazard communication program which will be available to employees, designated representatives of employees, OSHA, and NIOSH. The program must cover labeling, MSDSs, and employee training. It must include:

- a list of hazardous substances which are present in each work area or the whole workplace
- methods to inform employees of the hazards of nonroutine tasks, such as cleaning reaction vessels
- methods to inform employees of the hazards of chemicals in unlabeled pipes
- methods to inform onsite contractors and their employees of potential hazards

Labeling

Containers (bags, barrels, boxes, cans, drums, reaction vessels, storage tanks, bottles, etc.) of hazardous chemicals must be labeled by the manufacturer, importer, or distributor with:

- an identifier
- hazard warnings
- the importer or distributor's name and address

Employers shall ensure that each hazardous chemical container in the workplace is labeled with:

- an identifier
- hazard warnings

Labels may use a chemical or a common name, trade name, or code name. The hazard warning must describe the specific hazard (such as "This chemical causes liver damage"). Vague phrases such as "warning" or "danger" are not enough. *Portable containers* are not required to have labels as long as the chemicals are transferred from labeled containers, and the portable container is intended only for the immediate use of the employee who performs the transfer.

Labeling is not required for *pipes,* but written information on the content of unlabeled pipes must be available in the written program and the employee training program.

Material Safety Data Sheets

Employers are required to obtain and keep an MSDS for each hazardous substance at their facility. Manufacturers, importers, and distributors must forward an MSDS with the first shipment to an employer. MSDS copies must also be made available during each work shift to employees in their work areas, as well as to designated employee representatives, OSHA, and NIOSH.

The MSDS must include at least the following information:

- chemical *and* common name of the hazardous ingredients if they are greater than 1% of the mixture *or,* for potential cancer-causing substances, greater than 0.1% of the mixture
- physical and chemical characteristics (vapor pressure, flash point)
- physical hazards (including potential for fire, explosion, reactivity)
- health hazards, including signs and symptoms of exposure and any medical conditions made worse by exposure
- primary routes of entry (that is, breathing, skin contact, or swallowing)
- potential of the substance to cause cancer
- legal (OSHA) and advisory (ACGIH or other) exposure limits
- precautions for safe handling and use
- control measures, such as engineering controls, work practices, or personal protective equipment
- emergency and first aid procedures

- date prepared
- name, address, and telephone number of the chemical manufacturer, importer, or other responsible party preparing or distributing the MSDS

New information must be added to the MSDS within three months of when the manufacturer, importer, or employer became aware of it. If no relevant information is found for a given category on the MSDS, it shall be marked "No applicable information." Spaces must not be left blank. A sample form letter to request MSDSs is contained in Appendix B.

Employee Training

All employees covered by the Standard must be given training on chemical hazards in their work area on initial assignment and when new hazards are introduced. (See "Employee Education and Training," Chapter 4, Section 5.)
Training shall at least include:

- the requirements of the OSHA Standard
- any operations in the employee's work area where hazardous chemicals exist
- methods to detect the presence or release of a hazardous chemical in the work area (such as monitoring devices or odor)
- the physical and health hazards of the chemicals in the work area
- safe work practices, emergency procedures, and personal protective equipment
- the location, availability, and details of the employer's written hazard communication program, including the required lists of hazardous chemicals and MSDSs

Trade Secrets

A chemical manufacturer, importer, or employer may keep the chemical name and registry number off an MSDS if it is a trade secret. Information on the chemical's properties and effects, however, must still be disclosed, and the MSDS must indicate that the specific chemical identity is being withheld as a trade secret. In addition, the identity must be made available to health professionals in emergencies.

Enforcement

Violations of this Standard are handled in a similar way to other violations of OSHA rules and regulations. (See "Violations" and "Citations/Penalties," Section 1.) Employees can file a complaint with OSHA.

A complete copy of the Standard and a pamphlet describing it are available through the local area OSHA office. (See Appendix B.)

3. NEW JERSEY RIGHT TO KNOW

State Right to Know Laws

Due to the delays in the development of a federal labeling standard, proponents of such standards began, in the late 1970s, proposing state laws requiring chemical labeling. As of January 1, 1987, 20 state Right to Know (RTK) laws had been passed in Alaska, California, Connecticut, Illinois, Maine, Maryland, Massachusetts, Michigan, Minnesota, New Hampshire, New Jersey, New Mexico, New York, Oregon, Pennsylvania, Rhode Island, Vermont, Washington, West Virginia, and Wisconsin. Various cities, including Cincinnati, Philadelphia, and Santa Monica, have also enacted right-to-know legislation. The content and scope of the state and local laws vary widely. For a summary of employer's responsibilities under the New Jersey Right to Know law, see Appendix B.

New Jersey's Worker and Community Right to Know Act

On June 23, 1983, following nearly a year and a half of proposals, debates, public hearings, and numerous amendments to the bill, the New Jersey State Senate passed S.1670, and the companion bill in the State Assembly, A.3318, was passed the following Monday, June 27, 1983. New Jersey Governor Thomas H. Kean signed the Worker and Community Right to Know Act into law on August 29, 1983. Many provisions of the law took effect one year following enactment, on August 29, 1984.

Employers Covered

The original state law covered manufacturers (Standard Industrial Classification [SIC] Codes 20–39), state and local governments, and certain non-manufacturers designated by the Legislature as potential users of hazardous substances. A U.S. District Court ruling on January 3, 1985 (see "Court Update," below) exempted all manufacturers from coverage under the entire state law, ruling that those businesses were already covered by the federal OSHA Hazard Communication Standard. However, a U.S. Court of Appeals ruled on October 10, 1985 that manufacturing employers must comply with the *community* provisions of the state law. A later decision on August 24, 1987 extended the OSHA Hazard Communication Standard to cover non-manufacturers as well as manufacturers. Nonmanufacturers are now exempt from coverage under *worker* provisions of the state Right to Know law; however, they must comply with the *community* provisions.

Also, on January 21, 1986, Senate bill S.3435 was signed into law deleting certain SIC Codes originally covered by the state law, and adding new codes for coverage. No longer covered are travel agencies, car rental agencies, car washes, doctor's and dentist's offices, private nursing homes, medical and dental laboratories, and other nonmanufacturers. New employers included under the law are fuel oil dealers, landscaping and lawn care services, gas stations, dry cleaning plants, industrial launderers, and others. For an up-to-date list of covered employer categories, see Table 2.1.

Research and Development Laboratories

Laboratories that believe they qualify for a Research and Development (R&D) exemption from certain provisions of the law must submit an application for exemption to the New Jersey Department of Environmental Protection (DEP). This request will be reviewed and granted or denied based on the information supplied in the application. Approved research and development laboratories are exempt from survey reporting requirements. However, R&D laboratories operated by public employees still must provide education and training for their employees, maintain Hazardous Substance Fact Sheets, and set up a communications program with the local fire department. They may label containers using a code or number system as long as the employee can use the code or number to easily obtain the chemical name, Chemical Abstracts Service (CAS) number, trade secret registry number (if applicable), and a Hazardous Substance Fact Sheet for the substance.

Court Update: Legal Challenges to the New Jersey Right to Know Law and OSHA Hazard Communication Standard

Following the adoption of OSHA's Hazard Communication Standard in November, 1983, the states of New York, New Jersey, and Connecticut, which had passed state Right to Know laws, and several unions and public interest groups sued to defend the state laws against federal preemption.

In August 1984, the New Jersey Business and Industry Association, the New Jersey Chamber of Commerce, and the Chemical Industry Council sued to challenge the constitutionality of the new state Right to Know law. They argued that the federal OSHA Hazard Communication Standard explicitly preempted state Right to Know laws.

On January 3, 1985, the U.S. District Court (in Newark) ruled that the OSHA Standard completely preempted the New Jersey law (both worker and community provisions) as it applied to manufacturing operations. However, the state law remained fully intact for all *other* employers within its coverage. In addition, the court sustained the part of the act that established a Special Health Hazard Substance List which includes carcinogens, teratogens, highly flammable chemicals, etc., for which no trade secret claims could be made. Industry groups had argued that the act unconstitutionally took trade secrets from them without compensation. The judge ruled that, at least for this

Table 2.1   Standard Industrial Classifications—Major and Industrial Groups Covered Under the New Jersey Right to Know Law

SIC	Description
07	Agricultural Services
0782	Lawn and garden services[1]
20–39	Manufacturing Establishments (Entire Major Groups)[2]
20	Food and kindred products
21	Tobacco manufacturing
22	Textile mill products
23	Apparel and other textile products
24	Lumber and wood products
25	Furniture and fixtures
26	Paper and allied products
27	Printing and publishing
28	Chemicals and allied products
29	Petroleum and coal products
30	Rubber and miscellaneous plastic products
31	Leather and leather products
32	Stone, clay, and glass products
33	Primary metal industries
34	Fabricated metal products
35	Machinery, except electrical
36	Electrical and electronic equipment
37	Transportation equipment
38	Instruments and related products
39	Miscellaneous manufacturing industries
45	Transportation by Air
4511	Air transportation, certificated carriers[1]
4582	Airports and flying fields[1]
4583	Airport terminal services[1]

Table 2.1, continued

SIC	Description
46	Pipelines, Except Natural Gas (Entire Major Group)
47	Transportation Services
4712	Freight forwarding
4722	Arrangement of passenger transportation[3]
4723	Arrangement of freight and cargo transportation[3]
4742	Rental of railroad cars with care of lading
4743	Rental of railroad cars without care of lading
4782	Inspection and weighing services connected with transportation
4783	Packing and crating
4784	Fixed facilities for handling motor vehicle transportation, not elsewhere classified
4789	Services incidental to transportation, not elsewhere classified
48	Communication
4811	Telephone communication (wire or radio)
4821	Telegraph communication (wire or radio)
4832	Radio broadcasting[3]
4833	Television broadcasting[3]
4899	Communication services, not elsewhere classified[3]
49	Electric, Gas, and Sanitary Services (Entire Major Group)
50	Wholesale Trade—Durable Goods
5085	Machinery, equipment, and supplies—industrial[1]
5087	Machinery, equipment, and supplies—service establishments[1]
5093	Miscellaneous durable goods—scrap and waste[1]
51	Wholesale Trade—Nondurable Goods
5111	Printing and writing paper[3]
5112	Stationery supplies[3]
5113	Industrial and personal service paper[3]
5122	Drugs, drug proprietaries, and druggists' sundries
5133	Piece goods (woven fabrics)[3]

Table 2.1, continued

SIC	Description
5134	Notions and other dry goods[3]
5136	Men's and boys' clothing and furnishings[3]
5137	Women's, children's, and infants' clothing and accessories[3]
5139	Footwear[3]
5141	Groceries, general line[3]
5142	Frozen foods[3]
5143	Dairy products[3]
5144	Poultry and poultry products[3]
5145	Confectionery[3]
5146	Fish and seafoods[3]
5147	Meats and meat products[3]
5148	Fresh fruits and vegetables[3]
5149	Groceries and related products, not elsewhere classified[3]
5152	Cotton[3]
5153	Grain[3]
5154	Livestock[3]
5159	Farm product raw materials, not elsewhere classified[3]
5161	Chemicals and allied products
5171	Petroleum bulk stations and terminals
5172	Petroleum and petroleum product wholesalers, except bulk stations and terminals
5181	Beer and ale
5182	Wines and distilled alcoholic beverages
5191	Farm supplies
5194	Tobacco and tobacco products
5198	Paints, varnishes, and supplies
5199	Nondurable goods, not elsewhere classified
55	Automobile Dealers and Gasoline Service Stations
5511	Motor vehicle dealers (new and used)[1]

Table 2.1, continued

SIC	Description
5521	Motor vehicle dealers (used only)[1]
5541	Gasoline service stations—retail[1]
72	**Personal Services**
7216	Dry cleaning plants, except rug cleaning[1]
7217	Carpet and upholstery cleaning[1]
7218	Industrial launderers[1]
73	**Business Services**
7397	Commercial testing laboratories
75	**Automotive Repair, Services, and Garages**
7512	Passenger car rental and leasing, without drivers[3]
7513	Truck rental and leasing, without drivers[3]
7519	Utility trailer and recreational vehicle rental[3]
7523	Parking lots[3]
7525	Parking structures[3]
7531	Top and body repair shops
7534	Tire retreading and repair shops
7535	Paint shops
7538	General automotive repair shops
7539	Automotive repair shops, not elsewhere classified
7542	Car washes[3]
7549	Automotive services, except repair and car washes[3]
76	**Miscellaneous Repair Services**
7622	Radio and television repair shops[3]
7623	Refrigeration and air conditioning service and repair shops[3]
7629	Electrical and electronic repair shops, not elsewhere classified[3]
7631	Watch, clock, and jewelry repair[3]
7641	Reupholstery and furniture repair[3]
7692	Welding repair

Table 2.1, continued

SIC	Description
7694	Armature rewinding shops[3]
7699	Repair shops and related services, not elsewhere classified[3]
80	Health Services
8011	Offices of physicians[3]
8021	Offices of dentists[3]
8031	Offices of osteopathic physicians[3]
8041	Offices of chiropractors[3]
8042	Offices of optometrists[3]
8049	Offices of health practitioners, not elsewhere classified[3]
8051	Skilled nursing care facilities[3]
8059	Nursing and personal care facilities, not elsewhere classified[3]
8062	General medical and surgical hospitals
8063	Psychiatric hospitals
8069	Specialty hospitals, except psychiatric
8071	Medical laboratories[3]
8072	Dental laboratories[3]
8081	Outpatient care facilities[3]
8091	Health and allied services, not elsewhere classified[3]
82	Educational Services
8211	Elementary and secondary schools
8221	Colleges, universities, and professional schools
8222	Junior colleges and technical institutes
8231	Libraries and information centers[3]
8241	Correspondence schools[3]
8243	Data processing schools[3]
8244	Business and secretarial schools[3]
8249	Vocational schools, except vocational high schools, not elsewhere classified
8299	Schools and educational services, not elsewhere classified[3]

Table 2.1, continued

SIC	Description
84	Museums, Art Galleries, Botanical, and Zoological Gardens
8411	Museums and art galleries[3]
8421	Arboreta, botanical, zoological gardens[3]
91	Executive, Legislative, and General State, County, and Local Governments[4]

[1]Added for coverage by PL 1985, c. 543; N.J.S.A. 34:5A-3.
[2]U.S. Court of Appeals decision (10/10/85) reinstated NJDEP's authority to survey businesses in the manufacturing sector.
[3]Deleted from coverage by PL 1985, c. 543; N.J.S.A. 34:5A-3.
[4]This designation is given by NJDEP and NJDOH rather than listing all the separate categories.

special list, trade secret protection must give way to public health and safety concerns.

On May 24, 1985, the U.S. Third Circuit Court of Appeals (in Philadelphia) ruled that OSHA should extend its Standard to cover nonmanufacturing industries (construction, transportation, utilities, hospitals, auto repair, etc.). The Court also ruled that the OSHA Standard only preempted state laws for those industries covered by the OSHA Standard.

The Court also sent back the trade secret section of OSHA's Standard for reconsideration. It found that the Standard had given broader trade secret protection to employers than state law usually allows. For example, the Court will not allow OSHA to classify as a trade secret chemical identity information that is easily discoverable through reverse engineering. After failing to take expeditious action on the Third Circuit Court's order, OSHA was hit with a court order to extend the Standard to cover *non*manufacturing employers by July 29, 1987, so that the Standard now also applies to non-manufacturers.

On October 10, 1985, the U.S. Third Circuit Court of Appeals overturned part of the lower court's original ruling. The Court ruled that *all* covered New Jersey employers—including manufacturers—must comply with the *community* portion of the state Right to Know law. The Court did not make a decision on the issue of labeling "environmentally hazardous substances" in the manufacturing sector. They sent this issue back to the U.S. District Court in Newark for further arguments and a decision. The issue to be decided has been expanded to include universal labeling. Now, with the expansion of the OSHA Standard, the discussion will apply to both the manufacturing and nonmanufacturing sectors.

Another major impact on the operations of the New Jersey Right to Know law is the passage of new federal legislation. On October 17, 1986, the

Superfund Amendments and Reauthorization Act of 1986 (SARA) was signed into law by the President. Title III of SARA (Emergency Planning and Community Right-to-Know) requires emergency prevention and response planning and release notification by all employers who use, produce, or store extremely hazardous substances (designated by the EPA), as well as inventory reporting and toxic release reporting by all employers covered by the OSHA Hazard Communication Standard. The Standard covers the manufacturing sector (SIC Codes 20–39) and the nonmanufacturing sector. New Jersey has a strong commitment to continuing its Right to Know program while meeting requirements of the federal statute. Towards that end, the state has been meeting with EPA and discussing how the two programs can be adapted into a single operation.

For further updates on the status of this legislation, contact The Right to Know Project at the New Jersey State Department of Health (DOH).

Funding

Employers with 25 or fewer employees have to pay a minimum fee of $50 to the New Jersey Department of Labor. The fee for employers of more than 25 people is $2 per employee. This money is placed in the Right to Know Fund and pays the program's operation costs. According to the January 21, 1986 amendment to the law, if an employer reports that no hazardous substances are present at the facility, the employer will be entitled to a refund. Public employers do not have to pay the fee.

Employer Responsibilities

Each employer has six areas of responsibility under the act:

1. completion of Right to Know Survey*
2. completion of additional survey*—Title III, Superfund Amendments and Reauthorization Act
3. maintaining health records
4. making information available to employees
5. education and training
6. labeling

*These surveys are combined for all employers except public employers.

New Jersey Right to Know Survey. Each covered employer must complete a New Jersey Right to Know Survey for his/her work facility (this survey replaces the three separate surveys from the Departments of Health and Environmental Protection that have been used in the past). This includes manufacturers and nonmanufacturers, since the RTK survey includes the DEP's Environmental and Emergency Services Information Surveys required by the community Right to Know provisions of the law. On the survey, employers list substances from the New Jersey Right to Know Hazardous Substance List present at their facility. This list is composed of 2200 individual substances, 500 generic chemical categories, and 450 synonyms of chemicals which are potentially harmful to workers or the public's health and to the environment. Information on these hazardous materials is also necessary for emergency response situations. Surveys must be completed within 90 days. A due date is assigned to each survey and is printed on the upper right-hand corner of the mailing label. RTK county lead agencies and DOH will provide completed RTK Surveys and Hazardous Substance Fact Sheets to employees and the general public upon request, as will the DEP (surveys only).

The employer must keep a copy of the survey on file at the workplace, and send copies to the local police and fire departments, the RTK county lead agency, and to the DEP (DEP, in turn, will forward a copy to DOH). The DOH will automatically send Hazardous Substance Fact Sheets for all the substances listed on the survey to the employer.

It is not necessary to include substances which are on the New Jersey Right to Know Hazardous Substances List on the New Jersey Right to Know Survey, or to label them, if one of the following conditions is met:

- The article containing the substance is present in solid form which does not pose any health hazard to an employee. An article (such as an ash tray) is a manufactured item that (1) is formed to a specific shape or design during manufacture, (2) has end use function(s) dependent in whole or in part upon its shape or design during end use, and (3) does not release, or otherwise result in exposure to, a hazardous chemical under normal conditions of use.
- The concentration is less than 0.1% of a mixture (for carcinogens, mutagens, or teratogens coded as a Special Health Hazard Substance).
- The concentration is less than 1% of a mixture (for all other substances; unless the substance is present in a total amount of 500 pounds or more in a container at the facility).
- The substance is present in the same form and concentration as a product, packaged for distribution and use by the general public, to which an employee's exposure during handling is not significantly greater than a consumer's exposure during the principal use of the toxic substance.

Survey under Title III, Superfund Amendments and Reauthorization Act. The federal regulation, Title III of the Superfund Amendments and Reauthorization Act (SARA) of 1986, provides for community Right to Know and emergency response planning. Under this legislation, manufacturers and nonmanufacturers are required to report on hazardous and toxic chemicals (See "SARA Title III," Section 4, below). In New Jersey, a survey combining the requirements of the New Jersey Right to Know law and Title

III will be sent to manufacturers and nonmanufacturers by the New Jersey Department of Environmental Protection.

Health records. Upon request, employers must provide DOH with copies of employee health and exposure records.

Making information available to employees (public agencies only). Employers of public agencies must keep Right to Know Survey forms, the Right to Know Hazardous Substance List, and Hazardous Substance Fact Sheets in a central file at their facility. DOH sends to employers posters that must be posted on bulletin boards readily accessible to employees. The employer must request the number of posters needed from DOH. Any employee or employee representative may request from the employer copies of the facility's Right to Know Survey and Hazardous Substance Fact Sheets, and are entitled to receive them within five working days of the request. Employers must provide the survey form and fact sheets in Spanish, if requested by Spanish-speaking employees. DOH will provide these translations.

Education and training (public agencies only). Effective 1985, education and training on the hazardous substances employees are exposed or potentially exposed to must be conducted annually for current employees. New or reassigned employees must receive education and training within one month of hiring or reassignment.

The DOH adopted regulations (N.J.A.C.8:59–6) on setting up education and training programs. In response to many questions from employers, the DOH also developed the "New Jersey Worker and Community Right to Know Act Education and Training Program Guide" for developing an employee education and training program. Both of these documents are available. (See Appendix B.) The education and training program should include the following subjects:

- the common methods used to recognize occupational health and safety hazards
- the common methods used to measure and evaluate employee exposure to hazardous substances
- the common methods used to prevent and control employee exposure to hazardous substances
- an employee's rights and the employer's responsibilities under the New Jersey Right to Know law
- an explanation of the health and safety hazards of the hazardous substances listed on the New Jersey Right to Know Survey

- hands-on training in emergency response methods and personal protective equipment for certain designated employees
- a walkthrough of areas where employees are exposed or potentially exposed to hazardous substances

In addition, employers must maintain certain written records of all education and training programs, including a list of course objectives, copies of written materials used, and a roster with signatures of participants.

In order to qualify to conduct the training, the instructor must either (1) have a bachelor's degree in industrial hygiene, environmental science, health education, chemistry, or a related field, or be a registered nurse, and understand the health risks associated with exposure to hazardous substances; or (2) have at least 30 hours of hazardous materials training and one year of experience supervising employees who handle hazardous substances or work with hazardous substances, and understand the health risks associated with exposure to hazardous substances.

Labeling (public agencies only). A container is any bottle, pipeline, bag, barrel, box, can, cylinder, drum, carton, vessel, vat, or stationary or mobile storage tank. *All* containers at the facility must be labeled with the chemical or common name of the contained substance and its Chemical Abstracts Service (CAS) number, or its trade secret registry number, whether or not they contain hazardous substances. Along with the hazardous substances, the top five ingredients of the contents of every container must be labeled. Containers must be labeled before opening or within five working days of the containers' arrival at the facility, whichever is sooner.

Facilities such as *laboratories* that receive containers with unknown contents for the purpose of analysis must label the containers as soon as contents are identified. Employers who have containers with unknown contents, where the substance's manufacturer is unknown or no longer in business, must label the containers as required by this act or by the federal Resource Conservation and Recovery Act (RCRA). (See Chapter 3, Section 1.) In this situation, the employer is responsible for analyzing the contents of the container and then labeling it. An employer must make a good faith effort within six months of receipt of a New Jersey Right to Know Survey to determine the chemical name or CAS number of a substance or the components of a mixture by requesting the Material Safety Data Sheets from the manufacturer or distributor, with following phone calls or letters if necessary. If unsuccessful, the employer must notify DEP and DOH by completing the form entitled "Substances and Mixtures with Unknown Component(s)" and returning it to DEP.

Process containers, which are those containers that are changed at least once per shift, do not have to be labeled. *Pipelines* must be labeled at the valve(s) where a substance enters the pipeline and at valves, outlets, vents, drains, and sample connections designed to allow the release of the substance from the pipeline. The valves, outlets, vents, and drains of pipelines which control the emission or discharge of any solid, liquid, semisolid, or gaseous waste material from a facility must also be labeled. Certain federal labeling laws are acceptable substitutes for Right to Know labeling.

Trade Secrets

If an employer believes that disclosing information on the New Jersey Right to Know Survey or that labeling containers with the common or chemical name of a substance will reveal a trade secret, a trade secret claim must be submitted with the New Jersey Right to Know Survey. The copy of the survey provided to employees, local and county agencies, and the public will not contain the trade secret information. DOH or DEP will make trade secret information available to physicians when it is needed for medical treatment or diagnosis.

An employer may not file a trade secret claim for any substance on the Special Health Hazard Substance List (SHHSL), except if it is in a concentration low enough to exempt it from the SHHSL.

Employee Rights (public employees only)

This law gives full protection to employees from recrimination by employers for exercising employee rights under the act. In certain instances, an employee has the right to initiate civil action on his/her own behalf against his/her employer for any violation of the provisions of the act. An employee also has the right to refuse to work with a substance on the New Jersey Right to Know Hazardous Substance List if the employer has not provided him/her with the requested information within five working days. The New Jersey Department of Labor (DOL) will investigate complaints about recrimination and DOH will investigate complaints about any violations of the law.

Enforcement and Penalties

Enforcement of the Right to Know Act is the responsibility of DOH, DEP, and DOL. These agencies have the right to enter an employer's facility during its normal operating hours to inspect for compliance with the provisions of the act. The DOH and DEP have the right to fine an employer through a civil administrative penalty, to issue a civil administrative order, and to bring a civil action.

For resources on New Jersey Right to Know, see Appendix B.

References

Higgins, T. and P. Landsbergis. *A Guide to the New Jersey Worker and Community Right to Know Act* (Occupational Safety and Health Education Center, Rutgers University, 1983).

4. SARA TITLE III—EMERGENCY PLANNING AND COMMUNITY RIGHT-TO-KNOW

SARA Title III—Emergency Planning and Community Right-to-Know (P.L. 99-499)—establishes requirements for federal, state, and local governments and industry regarding emergency planning, notification, and hazardous substances inventory reporting.

Each state is required to establish a State Emergency Response Commission (SERC) which has the responsibility for overseeing and directing the federal requirements for their respective states. In New Jersey, Governor's Executive Order No. 161 established the SERC.

On February 13, 1987, New Jersey Governor Thomas H. Kean signed Executive Order 161. The existing Governor's Advisory Council for Emergency Services, with additional members, was designated to serve as the New Jersey SERC. The following representatives comprised the Advisory Council:

- Attorney General
- Adjutant General, Department of Defense
- Commissioner, Department of Community Affairs
- Commissioner, Department of Environmental Protection
- Commissioner, Department of Transportation
- President, Board of Public Utilities

Executive Order 161 also added for membership:

- Commissioner, Department of Health
- Superintendent, Division of State Police

The Commission is co-chaired by the Superintendent of State Police and the Commissioner of the Department of Environmental Protection. The New Jersey State Police Office of Emergency Management is responsible for the requirements of Title III, Subpart A ("Emergency Planning and Notification"). The Department of Environmental Protection is responsible for the requirements of Title III, Subpart B ("Reporting Requirements").

In order to implement the requirements of SARA Title III, the SERC is directed to designate local emergency planning districts and each of those districts is to have a local emergency planning committee (LEPC). In New Jersey, a two-tier system of local committees has been designated. They are the 567 municipalities and the 21 counties. New Jersey has a total of 588 LEPCs. The local committees should have members from the following groups and disciplines:

- elected state and local officials
- law enforcement
- emergency management
- fire service
- first aid squads
- health care workers
- local environmental organizations
- hospitals
- transportation
- broadcast and print media
- community groups
- facilities covered by this legislation

Local planning committees should rely heavily on existing emergency management structure.

These local committees are charged with the responsibility for developing emergency plans for their respective communities. The bases for these plans are notifications received from industry reporting that they have Extremely Hazardous Substances at or above certain specified Threshold Planning Quantities onsite at their facilities.

The committees also have responsibilities for the hazardous substances inventory reports required under SARA Title III. Each LEPC must receive copies of MSDSs or lists of data sheets and inventory reports from facilities within their jurisdiction. They must manage the information, advertise availability, and provide public access.

Every facility in the state, regardless of SIC Code or number of employees, has an obligation to review the List of Extremely Hazardous Substances and notify the SERC if they are subject to emergency planning.

Hazardous substances inventory reporting was required only from manufacturers (SIC 20–39) in 1988. As of 1989, *all* private sector employers are required to complete hazardous substances inventory reports. The New Jersey Department of Environmental Protection has developed a survey form that meets both the federal and state reporting requirements. (See Appendix B.)

In addition, there are toxic release inventory reporting requirements that only manufacturers must meet. They must report from a list of 329 substances at thresholds of 75,000 pounds manufactured or processed or 10,000 pounds otherwise used for 1988; 50,000/10,000 for 1989; and 25,000/ 10,000 for 1990 and every year thereafter.

There are three major programmatic differences between the state and federal programs that will require amendment to the state law so that both programs can be conducted jointly, not separately. The Department of Environmental Protection needs the authority to establish *thresholds* for substances that must be reported; in order to coordinate with the federal program, coverage of *all* private sector SIC Codes should be included in the state law; and New Jersey should adopt the OSHA Hazard Communication Standard as the criteria for hazardous substances inventory reporting, like the federal program.

For information on resources for SARA Title III, see Appendix B.

5. SMOKING IN THE WORKPLACE

Due to substantial evidence that smoking is harmful to the health of smokers and nonsmokers alike, a change in people's attitudes toward smoking has taken place. Smoking is now being banned or restricted in certain areas, as laws are created to protect the health of the public.

Research shows that cigarette smoking is the major preventable cause of premature death and disability in the United States. Tobacco smoke, which contains hundreds of chemical compounds, contributes to the occurrence of heart disease, lung disease, and cancer. Pregnant women who smoke risk affecting not only their own health, but also the health of their babies.

Smoking in the workplace has become an issue for employers due to the effects smoking has on the work environment. Concerns to employers about smoking include higher absenteeism rates for smokers than nonsmokers, lost productivity and wages due to smoking-related illnesses and premature deaths, and higher health care and maintenance costs for individuals who smoke. Some of the effects of tobacco smoke on nonsmoking employees are annoyances from the smell of smoke; symptoms such as headaches, cough, and eye and throat irritations; and increased risks of lung diseases. Also, employees who may have health conditions such as allergies, lung disease, or heart disease may be adversely affected by secondhand smoke.

Smoking in the Industrial Setting

According to NIOSH Current Bulletin 31 ("Adverse Health Effects of Smoking and the Occupational Environment"), there are six different mechanisms by which smoking may act with physical and chemical agents in the workplace to harm health.

1. Certain toxic agents in tobacco products and/or smoke may also occur in the workplace, thus increasing exposure to the agent. An example of this is carbon monoxide, a component of cigarette smoke, which can also be found as a workplace contaminant, most often in combustion processes.

   Other chemicals found in tobacco which workers might be exposed to at their jobs include acetone, acrolein, aldehydes (e.g., formaldehyde), arsenic, cadmium, hydrogen cyanide, hydrogen sulfide, ketones, lead, methyl nitrite, nicotine, nitrogen dioxide, phenol, and polycyclic aromatic compounds.

2. Workplace chemicals may be transformed into more harmful agents by the heat generated by smoking. It is important to note the temperature of burning tobacco in a cigarette is approximately 875°C (1600°F).

3. Tobacco products, by becoming contaminated with toxic agents found in the workplace, may facilitate entry of the agent into the body by inhalation, ingestion, or skin absorption.

4. Smoking can add to the damaging biological effects which result from exposure to toxic chemicals found in the workplace. For example, combined worker exposure to chlorine and cigarette smoke can cause a more damaging biological effect than exposure to chlorine alone.

5. Smoking can interact with worker exposure to toxic materials found in the workplace, resulting in more severe health damage than anticipated from adding the separate influences of the occupational exposure and smoking. An example of this is the increased risk of lung cancer in asbestos workers who smoke compared to all other smokers, and even greater risk than nonsmokers not exposed to asbestos.

6. Studies have shown that smoking contributes to accidents in the workplace. It has been suggested that injuries attributable to smoking were caused by loss of attention, preoccupation of the hand for smoking, irritation of the eyes, and coughing. Smoking can also contribute to fire and explosions in occupational settings where flammable and explosive chemical agents are used. However, in many of these areas smoking is prohibited.

Legislation on Smoking in New Jersey

By prohibiting smoking in the workplace, the health risks for smokers and nonsmokers are reduced. Nonsmoking employees find the smoke-free environment pleasant, and for many smokers it is an incentive to quit smoking. Employers benefit by possible reductions in plant and equipment maintenance costs (due to less damage caused by cigarette smoke), along with the possible lowering of health, life, and disability insurance rates.

In light of these facts, the State of New Jersey has passed several laws restricting smoking in workplaces and public places. These laws regulate when and where, rather than whether, a smoker may legally smoke.

The legislation controlling smoking in the workplace took effect on March 1, 1986. Public Law 1985, Chapter 184 pertains to private employers of 50 or more people in one location. (For employers of less than 50 people, compliance with the law is voluntary.)

The law states that employers are responsible for developing and implementing a written smoking-control policy for the workplace. If an employer wants to ban smoking completely or limit smoking areas, the law allows for a gradual "smoking phase-out" period during the first year the policy is in effect. All smoking policies implemented after March 1, 1986, must go into effect immediately. Employees shall be given a copy of the policy upon request.

The smoking policy may prohibit smoking in the entire facility, or designate areas where smoking is prohibited or permitted. Employers must display signs in areas indicating whether smoking is prohibited or permitted.

In the event that DOH suspects that a company is not complying with the act, the employer will be notified in writing. The DOH will provide recommendations for the company, and the employer may request a conference. The conference serves to provide the employer with information from the health department that would enable the company to comply with the law. If any company continues to violate the provisions of the act, court proceedings may take place to enforce compliance.

While not required by law, development of the policy should involve a survey of employees to establish a sense of how many people smoke and if areas and times should be designated to permit smoking. A committee to develop the policy should consist of representatives from management and labor, smokers and nonsmokers.

There are special considerations for smoking-permitted areas. These would entail ventilation and location of designated areas, break times, and accessibility for employees. Worksites should be designated as nonsmoking.

Employees should be informed and consulted prior to the development of the policy and of the date it is to go into effect. Employers may offer smoking education to employees as well as smoking cessation/reduction programs for smokers. (See Chapter 4, Section 7, "Employee Health Promotion.")

For smoking information resources and smoking cessation programs, see Appendix B.

References

American Lung Association. "Second-Hand Smoke, Take a Look at the Facts," 1982.

Centers for Disease Control, U.S. Department of Health and Human Services/Public Health Service. "Current Intelligence Bulletin 31: Adverse Health Effects of Smoking and the Occupational Environment," February 5, 1979.

New Jersey. "An Act Controlling Smoking in Places of Employment and Supplementing Title 26 of the Revised Statutes" (P.L. 1985, Chapter 184), 1985.

New Jersey State Department of Health. "Guidelines for Establishing a Policy for Controlling Smoking in the Workplace," 1986.

U.S. Department of Health and Human Services/Public Health Service, Office on Smoking and Health. "State and Local Programs on Smoking and Health," 1986.

CHAPTER 3

Environmental Legislation

William Goldfarb, JD, PhD

1. OVERVIEW

This chapter will provide an overview of federal and state environmental protection statutes that might affect a manager of a small industry, especially in New Jersey. Merely reading this chapter, however, will not substitute for consultation with a competent attorney. First, the depth in which these statutes can be discussed in this book is limited by space considerations. For example, the statutes themselves are described in detail but the important administrative regulations that clarify and implement the statutes can only be identified. Second, other requirements imposed by environmental law, in addition to these statutes, may also apply to a small industry. Three examples are common law liability rules, municipal ordinances, and tax codes. Third, the law may have changed between the time this chapter was written and the time it is read. But this chapter can substantially help a manager by warning him or her of potential environmental compliance problems before an enforcement situation arises. In other words, this chapter can alert a manager so that he/she can call in an attorney before it is too late.

An excellent means of determining the environmental regulatory exposure of an industry is to commission an "environmental audit." In an environmental audit, a team of engineers and attorneys, retained by the company, analyzes the company's site, process, waste disposal methods, regulatory requirements, and compliance with existing permits. A successful environmental audit can produce a least-cost plan for complying with governmental regulations and insulating the manager from liability. In addition, the environmental audit may be structured so that its results are protected from requests for disclosure. If a manager discovers that one of the statutes described in this chapter appears to cover his or her plant, following up with an environmental audit is urged.

This chapter will begin with a discussion of seven federal statutory provisions that frequently have an impact on small industry. Then, three New Jersey statutes will be analyzed. Because New Jersey statutes often serve as models for lawmakers in other states, readers outside New Jersey will also find this section useful. For further information about the federal statutes, a list of references is provided at the end of the chapter.

2. FEDERAL STATUTES

The Clean Water Act

The Clean Water Act (CWA) (33 USC sec. 1251 *et seq.*) declares that any "discharge" of a "pollutant" by a "point source" to "waters of the United States" is illegal without a discharge permit issued by either the EPA or a state that has been delegated authority to administer the permit program. New Jersey is one of approximately 40 states that are administering discharge permits. The permitting agency in New Jersey is the New Jersey Department of Environmental Protection (DEP).

A "discharge of a pollutant" is defined by the CWA as any addition of any substance, including heat, to a waterbody. A "point source" is a pipe, ditch, or other discrete conveyance, as opposed to a "nonpoint source," which consists of unconfined runoff. Only one variety of nonpoint source is regulated by the CWA, but this "ancillary pollution" exception is important for our purposes. Where an industrial site is a point source discharger of a toxic pollutant, the permitting authority may include in its discharge permit Best Management Practices to control plant site runoff, spillage or leaks, sludge or waste disposal, and drainage from raw material storage.

"Waters of the United States" is defined quite broadly in the CWA. This term includes drainage ditches, artificial waterways, mosquito canals, and wetlands. Although the CWA does not apply to groundwater discharges, New Jersey has gone further than federal law—as the CWA allows it to do—and requires a permit for point source groundwater discharges. Many "indirect discharges" (discharges into a sanitary sewer) also require state permits under New Jersey law. Discharges into storm sewers are also covered by the CWA and state law.

Point source dischargers, including indirect dischargers, are subject to technology-based "effluent limitations" (end-of-pipe discharge standards). Pollutants are divided into classes (conventional, toxic, nonconventional/

nontoxic, and heat). Each class of pollutants has a corresponding statutory pollution control criterion based on available technology. For example, dischargers of toxic pollutants must control their discharges based on a Best Available Technology Economically Achievable (BATEA) criterion. Variances based on economics and water quality are available to dischargers of nonconventionals/nontoxics and heat. Indirect dischargers may be entitled to "pretreatment credits" for toxic pollution that is treated by their sewage treatment plant. Discharges of certain pollutants, such as PCBs and various pesticides, are prohibited by the CWA.

Technology-based limitations broken down by classes of pollutants are uniform for all dischargers, wherever situated, producing a particular product by a particular process. For example, all electroplaters using the anodizing process must meet identical effluent limitations for heavy metals, whether they are located in New Jersey or Kansas. These uniform national technology-based effluent limitations may be found in federal effluent limitations regulations. Many small industries, however, do not conveniently fit within common industrial categories. Effluent limitations for them are set on an individual basis by the permitting authority applying its Best Professional Judgment based on the statutory criteria.

Dischargers on exceptionally clean or heavily polluted waterways might also have to meet effluent limitations more stringent than their technology-based limits. These water quality-based effluent limitations are based on an "anti-degradation" criterion for pristine waters and a "fishable-swimmable" criterion for other waters, although both criteria may be modified in unusual cases. Dischargers on "water quality-limited stretches," where the application of BATEA will still not meet the applicable water quality criteria, are awarded "wasteload allocations"—their proportionate shares of the pollution that the waterway can assimilate without exceeding the criteria. Meeting effluent limitations based on water quality may require installation of sophisticated control technology, raw material substitutions, process changes, or even product elimination.

Statutory Enforcement

Effluent limitations are included in discharge permits along with compliance schedules (generally spanning three years) and monitoring and reporting requirements. Permittees are required to submit to plant inspections (although a search warrant may be necessary), install and maintain monitoring equipment, and periodically report violations to the permitting authority. This "self-monitoring," supported by onsite "compliance monitoring," is the key to enforcement of the CWA. Thus, the CWA provides strict penalties for violations of monitoring and reporting requirements.

Most environmental protection statutes contain similar enforcement mechanisms. Thus, the CWA's enforcement provisions will be thoroughly explored as an example of statutory enforcement tools. During the remainder of this chapter, one can assume that enforcement of the other statutes discussed does not vary significantly from CWA enforcement unless indicated otherwise.

Environmental protection statutes establish "strict liability," i.e., liability without fault. And, of course, ignorance of the law is no defense to an enforcement action.

EPA's enforcement remedies are administrative compliance orders, administrative fines, civil suits, and criminal actions. These remedies need not be sought in any particular order, but in practice criminal sanctions are a last resort. State enforcement remedies must be adequate in order for the state to receive delegation of the permit program.

Class I administrative penalties may not exceed $10,000 per day of violation, with a maximum penalty of $25,000. Informal hearings are available in Class I penalty situations. Class II penalties may not exceed $10,000 per day of violation, with a maximum penalty of $125,000. There is an opportunity for a formal trial-type hearing in Class II penalty situations. EPA can determine in a specific case which class of penalties to apply.

Where the violation is more serious, EPA may bring a civil action for an injunction and a civil fine of up to $25,000 per day of violation. In determining the amount of a civil penalty, the court must consider the seriousness of the violation, any economic benefit to the violator, past violations, good faith efforts to comply, the economic impact of the penalty on the violator, and other relevant factors. The U.S. Supreme Court has held that an injunction need not be granted automatically when the CWA is violated. A court may "balance the equities" and determine whether an injunction would be in the public interest.

Any person, including a "responsible corporate officer," who "willfully or negligently" violates the CWA is subject to criminal prosecution. An unauthorized discharge of a hazardous substance into a sewer system, which foreseeably could cause personal injury or property damage, is a criminal offense. Negligent violations of the CWA are punishable by fines of between $2500 and $25,000 per day of violation, or by one year in jail, or both. These maximum sentences may be doubled for second and subsequent offenders. Knowing violations are punishable by fines of between $5000 and $50,000 per day of violation, or by up to three years in jail, or both. Once again, potential penalties are doubled for repeat offenders. A person who is found guilty of "knowing endangerment" (knowingly placing another person in imminent danger of death or serious bodily injury) is subject to fines of up to $250,000 and imprisonment for up to 15 years, with a possible doubling for repeat offenses. Persons making false statements in applications or reports, or tampering with monitoring devices, can be fined up to $10,000 and imprisoned for up to six months.

When neither the EPA nor DEP takes enforcement action against a permit violator, any citizen may sue a violator in federal court. If the citizen or citizens group wins the lawsuit (i.e., if the defendant is currently in violation of its permit) the judge can award an injunction and impose a civil fine on the defendant. In most cases, a losing defendant in a CWA citizen suit will be responsible for plaintiff's attorney and expert witness fees.

The Clean Air Act

Federal regulation of "stationary sources" (e.g., industrial plants) under the Clean Air Act (CAA, 42 USC sec. 7401 *et seq.*) is set in motion by EPA's establishment of National Ambient Air Quality Standards (NAAQS) for air pollutants that may endanger public health or welfare. The NAAQS for a particular pollutant must be set at a level of air quality that will protect public health with an adequate margin of safety. Factors other than public health (e.g., costs of compliance and technological feasibility of controls) cannot be taken into consideration in setting NAAQS. EPA has promulgated NAAQS for six pollutants: sulfur dioxide, particulate matter, carbon monoxide, nitrogen dioxide, ozone, and lead.

State Plans

NAAQS are not directly applicable to sources of these so-called "criteria pollutants." Instead, states must develop State Implementation Plans (SIPs) and have them approved by EPA. Each SIP must include a description of air quality by county, an inventory of sources that emit criteria pollutants, emissions limitations and compliance schedules to reduce emissions to levels that do not violate NAAQS, monitoring and enforcement requirements, enforcement strategies, and a permit program for review of new source construction to assure that new sources will not violate NAAQS.

SIPs should also contain state strategies for enforcing EPA-promulgated National Emissions Standards for Hazardous Air Pollutants (NESHAPS). Federal standards have been set for asbestos, beryllium, mercury, vinyl chloride, benzene, radionuclides, and arsenic. New Jersey regulates a number of additional hazardous air pollutants.

States have a good deal of flexibility in regulating sources of criteria pollutants. They may grant variances based on economic impact or technological feasibility as long as these variances do not cause violations of NAAQS. States may regulate similar sources uniformly or differently, depending on state policy. New Jersey requires virtually all sources of regulated pollutants to obtain emissions permits from DEP. Emissions limitations in these permits are based on state-of-the-art technology. Counties and municipalities play a major role in regulating emissions of particulate matter.

Air quality in states, or parts of them, is either better or worse than NAAQS. If it is worse, the area is called a "nonattainment area"; if better, it is called a Prevention of Significant Deterioration (PSD) area. New Jersey as a whole is a nonattainment area for ozone, and some of its urban centers are nonattainment areas for carbon monoxide.

Nonattainment Areas

New Jersey's status as a nonattainment area for ozone, and to a lesser extent carbon monoxide, is primarily due to the heavy concentration of automobiles on New Jersey roads. In order to obtain an extension to 1987 of the deadline for achieving the NAAQS for ozone and carbon monoxide, New Jersey established a vehicle inspection and maintenance program and proposed regulations regarding vapor recovery at service stations. However, these regulations were not promulgated and litigation ensued. Meanwhile, Congress has been considering another extension of the deadline. This area of air pollution control law is especially volatile, and developments in Washington, D.C. and in Trenton, New Jersey should be watched carefully, particularly by managers of businesses related to the use of automobiles. Will EPA use its authority to ban construction of new and modified sources in nonattainment states? Will federal highway funds and CAA grants be cut off if the states do not make significant progress toward attainment?

New sources in nonattainment areas are closely regulated under the CAA. As previously noted, a preconstruction review permit system must be included in each nonattainment SIP. Sources that will emit less than 100 tons per year of the nonattainment pollutant are exempt from the review requirement. In order to receive a permit, the applicant must show that (1) the new source will achieve the Lowest Achievable Emission Rate (LAER), (2) other sources owned or operated by the applicant in the state are in compliance with the SIP, and (3) a reduction in emissions (an "offset") has been obtained from existing sources—whether or not the applicant owns these sources—that would more than compensate for the new source's air pollution.

EPA's "offset policy" for new sources has a counterpart for existing sources, called the "bubble policy." A plant with several emissions points is treated as a whole, or as a bubble. Higher emissions than applicable regulations allow will be tolerated in exchange for lower emissions at another point. Only total emissions from the "bubble" will be considered.

Prevention of Significant Deterioration Areas

Many new sources in areas where the air is substantially cleaner than the NAAQS must also be permitted before construction. Under EPA's PSD area program, a source must undergo new source review if it falls within certain specified industrial categories and must emit more than 100 tons per year of a criteria pollutant. Potential sources outside these industrial categories would have to emit 250 tons of a criteria pollutant to fall under PSD review.

In order to receive a construction permit in a PSD area, the applicant must show that (1) the source will not cause a violation of NAAQS, (2) emissions of particulates or sulfur oxides will not exceed increments set by the CAA, and (3) the source will employ Best Available Control Technology (BACT) to control air pollutants. Visibility must be maintained in so-called "Class I Areas" bordering on resources such as national parks. In New Jersey, all major new sources that will emit criteria pollutants other than ozone and carbon monoxide must undergo preconstruction PSD review.

The Toxic Substances Control Act

The Toxic Substances Control Act (TSCA, 15 USC sec. 2601 *et seq.*) may be of major significance to a small manufacturer, processor, or importer of potentially toxic chemicals. Under TSCA, EPA is empowered to require the testing of new and existing chemicals that are potentially toxic, and to prohibit or place conditions on the manufacture, distribution, and usage of a chemical if it poses an unreasonable risk to human health or the environment.

TSCA regulates chemical substances and mixtures. It does not apply to pesticides (regulated by the Federal Insecticide, Fungicide, and Rodenticide Act), tobacco products, nuclear materials, or food additives, drugs, and cosmetics (regulated by the Federal Food, Drug, and Cosmetic Act).

Premanufacture Notice

The heart of TSCA is its requirement of a "premanufacture notice" (PMN) to EPA by any person who proposes to import, manufacture, or process a new chemical or to introduce a significant new use of an existing chemical. A chemical is considered to be "new" if it does not appear on the EPA's inventory of existing chemical substances. There is a limited exception to the PMN requirement for substances manufactured or imported in small quantities solely for research and development. Another exception applies to test marketing of a chemical.

The PMN submitter must produce information on the nature of the chemical (e.g., structure, trade name, amounts to be produced, and intended uses), the production process (e.g., number of workers and type of exposure, location of facilities, and environmental release data), and known effects (e.g., existing test data and literature citations).

Once a PMN is received, EPA must act within 180 days to regulate or prohibit the manufacture of the chemical. If EPA does not act within 180 days, the manufacturing process may begin. However, if on the basis of the PMN EPA finds that production, use, or disposal of the chemical "may present an unreasonable risk" to human health or the environment, EPA may promulgate "test rules" requiring further testing of health effects, environmental effects, and chemical fate. If these tests confirm the existence of an unreasonable risk, EPA may act to limit or prohibit production, use, or disposal. "Reasonableness" in TSCA consists of a "risk-benefit" analysis in which the health and environmental risk is weighed against the social and economic benefits of the chemical.

Existing Chemicals

As for existing chemicals, TSCA established an Interagency Testing Committee (ITC) that screens chemicals and recommends to EPA existing chemicals that are so potentially dangerous as to deserve further study. The ITC has developed several lists of potentially dangerous chemicals, including a "priority list" and a "suspicious list." Chemicals on these lists are subject to extensive monitoring and reporting requirements with regard to production, release, and exposure data. Each of these lists has its own set of requirements and exemptions, including small business exemptions. The manager of a small industry that manufactures or processes chemicals should procure accurate and current information about whether TSCA applies to his or her facility and, if so, what is required by way of monitoring and reporting.

All manufacturers and some processors and distributors must maintain records of and report to EPA on significant adverse health or environmental reactions caused by a substance or mixture that they handle. Only information actually submitted to the company must be recorded and reported; there is no obligation under TSCA to seek out this information.

All manufacturers, processors, and distributors must report any information that "reasonably supports the conclusion" that one of their substances or mixtures presents a substantial risk of injury to health or the environment. Managers may be subject to criminal liability if this duty is not fulfilled.

EPA has also been given broad powers under TSCA to order individual firms to submit health and safety studies in their possession.

When a report by a manufacturer, processor, or distributor indicates that the substance or mixture might present an unreasonable risk to human health or the environment, EPA may require further tests or act to limit manufacture, use, or disposal of the chemical.

Finally, TSCA establishes a framework for regulating certain extraordinarily dangerous chemicals, for example, asbestos, TRIS, and PCBs. TSCA is one of the federal statutes that could be relied upon to regulate products of genetic engineering.

The Resource Conservation and Recovery Act

The Resource Conservation and Recovery Act (RCRA, 40 USC sec. 6901 *et seq.*) applies to all industries that generate, transport, store, treat, or dispose of hazardous waste. Before 1984, EPA administratively exempted "small quantity generators" of less than 1000 kg/month of hazardous waste from RCRA's requirements. However, the 1984 RCRA amendments lowered the small quantity generator threshold to 100 kg/month, except for generators of acutely hazardous wastes who are regulated if they generate more than 1 kg/month of these wastes. Generators of less than 100 kg/month of hazardous waste are not required to comply with the manifest system; but they must still test their wastes to determine if they are acutely hazardous, and they must also dispose of their wastes onsite or at a permitted facility. Generators of between 100 and 1000 kg/month must comply with the manifest system, but are given some relief from recordkeeping and other requirements.

Hazardous Waste Management

RCRA is strictly a waste management statute. It does not apply to storage or use of hazardous raw materials or products unless a hazardous waste is generated. Moreover, discharges of hazardous waste into a sanitary sewer are exempt from RCRA.

RCRA's approach to hazardous waste management consists of four major elements:

- federal identification of hazardous wastes
- a manifest system for tracing hazardous wastes from generator to transporter to treatment, storage, or disposal facility
- federal minimum standards for hazardous waste treatment, storage, and disposal, enforced through a permit system
- state implementation of hazardous waste management programs at least equivalent to the federal program

Hazardous wastes are listed by chemical constituent and by waste stream. But if a waste does not appear on these lists it may still be hazardous, and it is the responsibility of each generator to test its waste according to EPA regulations. One such test, the so-called "EP Toxicity Test," simulates leaching into groundwater. Generators of hazardous wastes must also (1) keep records and report to EPA or an administering state, (2) initiate a manifest system "to assure that all such hazardous waste generated is designated for treatment, storage, or disposal in facilities for which a permit has been issued," and (3) properly label and containerize hazardous wastes delivered to transporters and treatment, storage, and disposal facilities. The duties of a transporter involve (1) recordkeeping and reporting, (2) accepting only properly labeled and containerized wastes, (3) complying with the manifest system, and, most important, (4) transporting all hazardous waste only to the permitted facility that the generator identifies on the manifest. The manifest system terminates with the receipt of the wastes by the owner or operator of the permitted facility and his/her notification to the generator. In New Jersey, copies of the manifest must also be sent to DEP by the generator and the owner or operator of the facility.

Performance Standards

Permits are not required of generators or transporters of hazardous wastes, but treatment, storage, or disposal of these wastes—including onsite activities of this nature—are prohibited except in accordance with a permit. In order to obtain and retain a permit, a facility operator must show that the facility meets EPA performance standards governing location, design, construction, operation, and maintenance. The applicable performance standards become permit conditions, in addition to the recordkeeping and reporting requirements. A facility operator must also comply with EPA regulations regarding preparedness and prevention, contingency plans and emergency procedures, closure and postclosure measures, and financial responsibility.

An important set of performance standards relates to groundwater protection. The facility permit must contain conditions to ensure that hazardous constituents from a regulated unit do not exceed certain concentration limits at the downgradient facility boundary for the active life of the facility and during postclosure if the site has not been decontaminated on closure. A "hazardous constituent" is a constituent on the RCRA list that has been detected in the underlying aquifer or "that is reasonably expected to be in or derived from waste contained in a regulated unit." However, the permitting authority may grant a variance if it "finds that the constituent is not capable of posing a substantial present or potential hazard to human health or the environment." "Concentration limits" are background levels, drinking water standards where they exist, or alternate concentration limits based on the variance criterion for a hazardous constituent. In other words, permit limitations are set on a site-specific basis. Moreover, a site is exempt from groundwater protection standards if the permitting authority finds "that there is no potential for migration of liquid from a regulated unit to the uppermost aquifer during the active life and the post-closure period."

All facilities treating, storing, or disposing of hazardous wastes must perform detection monitoring. EPA may exempt from monitoring requirements land disposal units that are both designed to prevent liquids from entering the unit and are equipped with multiple leak detection systems. Any wastes found beyond the actual waste management area are presumed to originate from the regulated unit. When a hazardous constituent is detected beyond the facility boundary, the owner or operator must take "corrective action," even beyond the boundary, to remove the waste constituents or treat them in place. Thus, RCRA contains a cleanup program that must be integrated with other cleanup programs to be discussed later in this chapter.

The 1984 RCRA amendments significantly strengthen RCRA's protection of groundwater. First, the land disposal of hazardous waste, including deep well injection, must be banned unless EPA determines that a particular method of land disposal will be consistent with protecting human health and the environment. A method of land disposal cannot be acceptable unless a petitioner demonstrates that there will be no migration from the land disposal unit for as long as the waste remains hazardous. If a disposability determination is not made within 66 months of the passage of the amendment, land disposal of the hazardous waste is automatically banned. Second, the landfilling of noncontainerized liquids is prohibited, and landfilling of containerized liquids must be minimized.

Third, facilities permitted under "interim status" must either install double liners, leachate collection systems, and groundwater monitoring devices or stop receiving hazardous wastes. Numerous exceptions and modifications are available, but impoundments with clay liners must close as storage impoundments. And finally, a new facility must have a double liner with leachate collection above and between the liners, and must also monitor groundwater. A variance from the double-liner requirement is available where the owner or operator can show that an alternative design, together with locational characteristics, is as effective in preventing migration of hazardous constituents to groundwater.

The 1984 amendments also establish a regulatory program to control underground storage tanks containing hazardous substances and petroleum products. Excluded, among other things, are farm and residential tanks storing motor fuels, heating oil tanks used for noncommercial purposes, septic tanks, regulated pipelines, surface impoundments, stormwater and wastewater collection systems, and flow-through process tanks. The regulatory program for existing tanks includes requirements for notification of the permitting authority by tank owners and operators; performance standards, leak detection systems, inventory control, and tank testing; recordkeeping and reporting; corrective action; financial responsibility; and closure. For new tanks, the requirements involve design, construction, installation, release detection, and compatibility with soils. States may be delegated authority to administer the underground storage tank program. New Jersey administers an active program that in some respects goes beyond federal law.

Section 311 of the Clean Water Act (Oil Spills)

Liability for oil spills is covered under four federal statutes. Section 311 of the CWA is the most comprehensive of these statutes. The others were enacted to control oil pollution in specific areas. They are The Outer Continental Shelf Lands Act Amendments, The Deepwater Port Act, and The Trans-Alaska Pipeline Authorization Act. These four statutes are similar in that they establish trust funds to pay for cleanup of oil spills. However, they differ in many ways, and Congress has been considering a comprehensive oil pollution control statute.

Section 311 of the CWA deals with oil and hazardous substance liability. However, with regard to hazardous substances it must be read together with the Comprehensive Environmental Response, Compensation and Liability Act (CERCLA).

The coverage of Section 311 is limited to "spilling, leaking, pumping, pouring, emitting, emptying, or dumping" oil and hazardous substances from vessels, onshore facilities, and offshore facilities into fresh or marine waters of the United States seaward to the 200-mile limit. It does not cover ground-water discharges or point source discharges under CWA discharge permits. Section 311 is administered by EPA and the Coast Guard.

Requirements of Section 311

The first step in the 311 process is for EPA to determine "those quantities of oil and any hazardous substance the discharge of which may be harmful to the public health or welfare." In 1979, EPA promulgated regulations designating approximately 300 substances as hazardous and establishing a "reportable quantity" for each. Virtually all discharges of oil are reportable. Any person in charge of a vessel or an onshore or offshore facility is required to immediately notify the Coast Guard or EPA as soon as he or she has knowledge of the discharge of a reportable quantity of oil or a hazardous substance. Failure to notify is punishable by fines and imprisonment.

Apart from the notification requirement, discharge of a reportable quantity gives rise to responsibility for both penalties and the costs of removal. Whenever any oil or a hazardous substance is discharged, the Coast Guard and EPA are authorized to "remove or arrange for the removal of such oil or substance at any time" unless they determine that the removal will be done properly by the owner or operator. In addition to removal, the federal government may "act to mitigate the damage to the public health or welfare caused by the discharge."

Liability

An owner or operator is liable to reimburse the federal government for the actual costs of cleanup and mitigation. Limitations on liability differ among onshore facilities, offshore facilities, and vessels, and among dischargers of oil and hazardous substances. However, an owner or operator is liable for the full amount of the costs where the discharge "was the result of willful negligence or willful misconduct within the privity and knowledge" of the responsible person. Removal costs include restoration or replacement of natural resources damaged or destroyed by the discharge. Under Section 311, a discharger is exempt from cleanup liability only if he can prove that the discharge "was caused solely by (a) an act of God, (b) an act of war, (c) negligence on the part of the United States Government, or (d) an act or omission of a third party without regard to whether such an act or omission was or was not negligent, or any combination of the foregoing causes." This is often referred to as "strict liability" because the exercise of care is no defense.

Section 311 includes a revolving fund, financed by fine receipts and congressional appropriations, to support federal oil spill cleanup efforts where financially responsible parties that are willing to undertake cleanup cannot be located. The 311 fund may not be used to compensate "third party" spill victims, that is, owners of shorefront property, fishermen, and others damaged by spills.

Comprehensive Environmental Response, Compensation and Liability Act

The Comprehensive Environmental Response, Compensation and Liability Act (CERCLA, 42 USC sec. 1901 *et seq.*), also known as the "Superfund," applies to "releases" of "hazardous substances" from "facilities" and vessels. The term "release" is extraordinarily comprehensive, covering "any spilling, leaking, pumping, pouring, emitting, emptying, discharging, injecting, escaping, leaching, dumping, or depositing into the environment." Abandoned hazardous waste containers are also releases. Unlike Section 311, CERCLA applies to groundwater discharges. "Release" does not include specified discharges of nuclear materials, workplace emissions, most engine exhausts, and normal fertilizer applications. Moreover, a release of a pollutant under a federal permit, a release from the application of a registered pesticide, and a release mentioned in an environmental impact statement are exempt from CERCLA.

"Hazardous substances" are any substances listed as toxic or hazardous under any federal pollution control statute. A "facility" is:

- any building, structure, installation equipment, pipe or pipeline (including any pipe into a sewer or publicly owned treatment works), well, pit, pond, lagoon, impoundment, ditch, landfill, storage container, motor vehicle, rolling stock, or aircraft
- any site or area where a hazardous substance has been deposited, stored, disposed of, or placed, or otherwise come to be located, excluding any consumer product in use or any vessel

This definition includes abandoned hazardous waste sites.

CERCLA contains spill notification provisions that are similar to those of Section 311. Moreover, owners and operators of vessels or facilities handling hazardous wastes must show proof of financial responsibility. Unless the Coast Guard and EPA determine that the person responsible for the spill will clean it up, the federal agencies may arrange for pollution removal and remedial operations whenever "any hazardous substance is released or there is a substantial threat of such release into the environment."

Financing Response Activities

CERCLA establishes an $8.5 billion fund over five years to finance Superfund response activities. The sources of revenue for the fund are:

- a petroleum tax of 8.2 cents per barrel for domestic crude oil and 11.7 cents per barrel for imported petroleum products, including imported crude oil—expected to raise $2.75 billion
- a chemical feedstock tax—expected to raise $1.4 billion
- an environmental tax based on corporate minimum taxable income, at a 12% rate over $2.5 billion
- general revenues of $1.25 billion
- fund interest and cost recoveries—expected to raise $600 million

When a release has occurred, EPA can use fund monies to clean up the site and then proceed against responsible parties for reimbursement, or move against responsible parties in the first instance. CERCLA imposes strict "joint and several" (recovery may be obtained from one or all) liability against (1) current owners or operators of facilities, (2) owners or operators at the time the hazardous substances were released, and (3) generators and transporters of the hazardous substances that were ultimately released by the facility. Only innocent purchasers who have made reasonable investigations are insulated from cleanup liability. CERCLA contains dollar limitations on liability, but these ceilings are considerably higher than those under Section 311.

Responsible parties may also be required to pay for "restoring, replacing, rehabilitating, or acquiring the substantial equivalent of damaged natural resources," but fund monies cannot be used for this purpose. Only natural resources belonging to, managed by, or protected by a state or the federal government are eligible. Moreover, there is no cleanup liability for resource damage occurring "wholly before" CERCLA's enactment, or for losses arising from "long-term exposure to ambient concentrations of air pollutants from multiple or diffuse sources" (e.g., acid precipitation). Like Section 311, CERCLA does not provide for compensating private victims of hazardous substance releases.

Short-term removal actions may be financed exclusively by the fund, but states must bear at least 10% of the cleanup costs and all future site maintenance expenses where long-term remedial activities are necessary. Where a site is owned or operated by a state or municipal government, the state must pay 50% of all costs of remedial operations. Fund monies may not be used for remedial actions where a state does not have adequate facilities for disposing of all its hazardous wastes for 20 years.

Determining Priorities

Federal hazardous substances response must, to the greatest extent possible, be consistent with the National Contingency Plan (NCP) begun under Section 311 and updated under CERCLA. The NCP includes preferable removal methods, cost-effectiveness criteria, the roles of governmental units, and criteria for determining cleanup priorities. Perhaps most important, the NCP shall "list national priorities among the known releases or threatened releases throughout the United States," subject to annual revision. This is known as the "National Priorities List" (NPL).

As an adjunct to the regulation of underground storage tanks under RCRA, the 1986 CERCLA amendments establish an "Underground Storage Tank Trust Fund" financed by a tax of 0.1 cent per gallon on petroleum fuels. EPA or a state may require an owner or operator of a tank to undertake corrective action if the action will be performed properly and promptly, or the agency may undertake the corrective action if necessary to protect human health and the environment. Fund monies may be used for financing corrective actions, enforcement, or cost-recovery proceedings.

CERCLA amendments enacted in 1986 have a significant bearing on how quickly, and to what extent, NPL sites will be cleaned up. Remedial investigation/feasibility studies for facilities on the NPL must be commenced at the following rate: 275 within three years, an additional 175 within four years, and an additional 200 within five years, for a total of 650 by 1991.

The amendments require EPA to select, to the maximum extent practicable, remedial actions that utilize permanent solutions and alternative treatment technologies or resource recovery technologies. A preference is established for remedial actions that utilize treatment to permanently and significantly reduce the volume, toxicity, or mobility of hazardous substances. Offsite transport and disposal without treatment is the least preferred option where practicable treatment technologies are available. If the selected remedy does not achieve the preference for treatment, EPA must publish an explanation. For onsite remedial actions, the amendments require attainment of "legally applicable or relevant and appropriate federal and state standards, requirements, criteria, or limitations" (ARARs), unless such requirements are waived. Maximum contaminant level goals under the Safe Drinking Water Act and water quality standards under the Clean Water Act must be met where relevant and appropriate.

SARA Title III: Emergency Planning and Community Right-to-Know Act of 1986

Title III of the 1986 CERCLA amendments (also known as the Superfund Amendments and Reauthorization Act [SARA]) establishes a new regulatory program that will require disclosure of information to workers and the general public about the dangers of hazardous chemicals, as well as development of emergency response plans for chemical emergencies.

Emergency Planning

Emergency response plans will be prepared by local emergency planning committees under the supervision of state emergency response commissions. Local committees are appointed by their state commissions; and each committee must include representation from a wide range of community groups, including owners and operators of facilities in the planning area. Plans for responding to chemical emergencies were to have been completed by 1988.

EPA has published a list of 402 extremely hazardous substances with threshold amounts for each chemical. Operators of facilities that use these listed chemicals in amounts exceeding the thresholds must (1) notify the state emergency response commission and (2) send a representative to participate in plan preparation with the local planning committee. Some of the threshold limits have been set as low as 2 pounds.

Any facility that releases one of these extremely hazardous chemicals in a reportable quantity under Section 311 and CERCLA must immediately report to the community emergency coordinator and the state commission. If no reportable quantity has been set for the particular substance released, then releases of more than 1 pound must be reported. Any facility that uses an extremely hazardous substance over its threshold limit must even report releases of substances that do not appear on the extremely hazardous substance list. A facility need not provide notification of a chemical release if the released substance remains within the confines of the facility.

Facilities that are required by the federal Occupational Safety and Health Act to have MSDSs for hazardous chemicals must submit a copy to the local committee, the local fire department, and the state commission. All MSDSs must be made available to the public. In addition, these facilities must prepare and submit detailed "emergency and hazardous chemical inventory forms" which describe amounts of hazardous chemicals present at the facility and their approximate locations.

Toxic Chemical Release

Section 313 of CERCLA requires certain facilities to submit to EPA and the host state an annual "Toxic Chemical Release Form" reporting on the total quantity of listed chemicals used onsite and released to the environment—either accidentally or intentionally—through discharges to air, water, and land. The deadline for submitting the first reports was July 1, 1988; reports must be submitted annually thereafter on July 1.

EPA has published a list of 309 chemicals and 20 categories of chemicals that trigger Section 313's reporting requirements. This list may be expanded by EPA at any time.

Facilities subject to Section 313 are facilities that have 10 or more full-time employees, and are within SIC Codes 20–39 inclusive, that manufacture, import, process, or otherwise use listed chemicals in amounts exceeding the following: (1) for facilities that manufacture, import, or process listed chemicals, 75,000 pounds during calendar year 1987, 50,000 pounds during 1988, and 25,000 pounds during 1989 and thereafter; and (2) for facilities that otherwise use listed chemicals, 10,000 pounds per year in 1987 and thereafter.

Toxic Chemical Release Forms must be made available to the public, subject to confidentiality claims. No new monitoring requirements are imposed on facilities by Section 313. If accurate data are unavailable, the Toxic Chemical Release Form may be based on an estimated mass balance.

Title III's Right-to-Know provisions do not preempt state laws such as the New Jersey Right to Know Act. (See Chapter 2, Section 3.)

3. NEW JERSEY STATUTES

Toxic Catastrophe Prevention Act

The Toxic Catastrophe Prevention Act (TCPA, N.J.S.A. 13:1K–19 *et seq.*) establishes an initial list of twelve "extraordinarily hazardous substances" and requires the New Jersey Department of Environmental Protection to develop and update an extraordinarily hazardous substance list. Each facility in the state that generates, stores, or handles one of these substances must register with DEP and submit extensive information about the facility's use of the substance. Some of the information required by the TCPA overlaps information required by Title III of CERCLA, but the TCPA requirements go beyond those of Title III.

If the facility already possesses a risk management program, DEP will review it and, if necessary, recommend changes or additions. If DEP and the owner and operator agree on changes or additions, they may enter into a consent agreement. If they cannot agree, DEP may issue an administrative order requiring these changes or proceed as follows.

Where a facility does not have a risk management program acceptable to DEP, the agency must develop an "Extraordinarily Hazardous Substance Risk Reduction Work Plan," which will form the basis of an "Extraordinarily Hazardous Substance Accident Risk Assessment" prepared for the facility by an independent consultant chosen by DEP. Based on the risk assessment, DEP must order the facility to undertake an "Extraordinarily Hazardous Substance Risk Reduction Plan" specifying steps to be taken by the facility and a compliance schedule for implementing them.

DEP must promulgate regulations for dealing with confidential information under TCPA, and develop a fee schedule so that the TCPA program can be self-supporting.

Spill Compensation and Control Act

State spill compensation and control statutes are important because (1) they impose liability where CERCLA does not (for example, for third-party damage), and (2) they provide extra money to finance the state's share of CERCLA cleanup, or to clean up sites that are not on the NPL. Approximately 23 states have statutory mechanisms to cover cleanup costs, and 13 states have reportedly enacted statutes allowing private citizens to recover for damages caused by hazardous substance spills.

The New Jersey Spill Compensation and Control Act (SCCA, N.J.S.A. 58:10–23.11 *et seq.*), for example, was enacted in 1976, four years before CERCLA. SCCA establishes a revolving New Jersey Spill Compensation Fund, financed by a tax on major facilities that refine, produce, store, handle, transfer, process, or transport oil or hazardous substances. A facility is "major" if it has total combined aboveground or buried storage capacity for hazardous substances of 200,000 gallons or more for nonpetroleum substances, or 20,000 gallons or more for petroleum-related substances. The fund may be used to remove actual discharges or threatened discharges if hazardous materials are dangerously stored or transported. Discharges that occurred prior to the act's passage may also be cleaned up with fund monies, but there are statutory limits on how much may be spent to clean up preenactment sites in any one year.

The fund holds any person who has discharged a hazardous substance or is in any way responsible for any hazardous substance removed by DEP strictly liable for all cleanup and removal costs and for all direct and indirect damages, no matter by whom they are sustained. Compensable damages include:

- costs of restoring and replacing natural resources
- costs of restoring, repairing, or replacing private property

- loss of income due to property or resource damage
- loss of tax revenues by state or local governments for one year due to property damage
- interest on loans taken out by a claimant pending fund payments

The SCCA does not mention personal injury costs, but would probably be construed to cover them.

Any person who has discharged a hazardous substance or is in any way responsible for any hazardous substance removed by DEP is strictly liable for all cleanup and removal costs. Where damage costs are concerned, there are dollar limitations similar to those in Section 311. The only defenses to liability are acts of war, sabotage, and acts of God. The SCCA has been held to apply retroactively, that is, to impose liability for preenactment discharges that are still a threat to the public.

When the fund has been utilized to pay cleanup costs and damages, the fund may then proceed against the discharger for reimbursement. Costs paid by the fund become a "superlien"—a first priority claim—against the property where the discharge occurred.

The SCCA prohibits all discharges of hazardous substances except in accordance with federal or state permits. Any unpermitted discharge of a hazardous substance must be reported to DEP, the municipality, and the local board of health. This aspect of the SCCA is similar to the release reporting requirement of CERCLA's Title III. However, the SCCA release notice is given to different organizations, and the SCCA list of hazardous substances is much broader than the Title III list of extremely hazardous substances.

Environmental Cleanup Responsibility Act

The Environmental Cleanup Responsibility Act (ECRA, N.J.S.A. 13:1K–6 et seq.) is becoming a model for other states. Massachusetts has enacted a similar statute and a number of other states are considering ECRA-type legislation.

Under ECRA, the owner or operator of an "industrial establishment" must notify DEP prior to closing, selling, or transferring its plant site. An "industrial establishment" is defined as an operation that generates, manufactures, refines, transports, treats, stores, handles, or disposes of hazardous substances onsite, and has a SIC number within Codes 22–39, 46–49, inclusive, or 51 or 76. Landfills and establishments engaged in the production and distribution of agricultural commodities are exempt from ECRA.

Before the closure, sale, or transfer can be legally finalized, the owner or operator must submit to DEP for approval either (1) a "negative declaration" that there has been no discharge of hazardous substances on the site or that the discharge has been cleaned up, or (2) a cleanup plan along with financial security guaranteeing the cleanup. If the premises are being transferred for approximately the same uses, DEP may permit deferral of the cleanup plan's implementation until the transferee closes, terminates, or transfers operations.

If the transferor does not comply with ECRA, the transferee may void the sale and recover damages. DEP may also void the transfer. Moreover, the transferor remains strictly liable for all cleanup costs and damages resulting from its failure to comply with ECRA. Because of the transferor's strict liability and the voidability of a transfer under ECRA, transferees and their lending institutions require compliance with the statute whether or not there has been a known use of hazardous substances on the site.

References

Arbuckle, J. G., N. S. Bryson, D. R. Case, C. T. Cherney, R. M. Hall, Jr., J. C. Martin, J. G. Miller, M. L. Miller, W. F. Pedersen, Jr., R. V. Randle, R. G. Stoll, T. F. P. Sullivan, and T. A. Vanderver, Jr. *Environmental Law Handbook,* 9th ed. (Rockville, MD: Government Institutes, Inc., 1987).

Goldfarb, W. *Water Law,* 2nd ed. (Chelsea, MI: Lewis Publishers, Inc., 1988).

Hayes, D. J., and C. B. MacKerron. *Superfund II: A New Mandate* (Washington, DC: Bureau of National Affairs, 1987).

Stever, D. W. *Law of Chemical Regulation and Hazardous Waste* (Clark Boardman Company, Ltd., 1987).

Selected Environmental Law Statutes (West Publishing Co., 1987).

CHAPTER 4

Employee Safety and Health and Industrial Hygiene

1. EMPLOYEE HEALTH

Sections 1 and 2 of this chapter were adapted from New Jersey Worker and Community Right to Know Act, Education and Training Program Guide, Right to Know Project, Occupational Disease Prevention and Information Program, New Jersey State Department of Health, 1986.

Most health care providers are not trained to determine if an illness or symptoms are related to chemical exposures at work. Thus, most physicians are not likely to ask a patient about potential exposures to hazardous substances at work. Information such as Material Safety Data Sheets and the New Jersey State Department of Health's Hazardous Substance Fact Sheets may be useful to health care providers. Employees who work with chemicals should be encouraged to provide their personal physician with a list of chemicals with which they work. Some physicians specialize in occupational medicine and are trained to recognize chemical exposures; there are about 1000 physicians certified in occupational medicine by the American Board of Preventive Medicine. Many other practitioners have gained substantial experience in this field. However, compared to other medical specialties, there are still only a few thousand physicians with expertise in occupational medicine. (For information on locating occupational medicine practices, see Chapter 5, Section 6.)

Medical Surveillance

Medical surveillance involves periodic examination of employees to detect subtle changes indicative of exposure, before disease has developed. Medical surveillance cannot be substituted for preventing exposure, but may be useful to assure that controls are effective. There are basically two types of job-related medical tests: disease monitoring tests and tests for toxic substances. Disease monitoring tests look for evidence that an employee has developed an occupational disease. These include chest X-rays, lung function tests, blood or urine tests for kidney or liver function, and EKGs to check the heart. Tests for toxic substances in the blood, breath, urine, hair, or other parts of the body are known as biological monitoring.

These two tests can benefit employees in the following ways:

- They can pick up health problems early, if done periodically, and allow time for correction of the hazard.
- They can detect change over time, to determine if health is getting worse. An initial measure of health, at the beginning of employment, establishes the baseline.
- They are useful for studies which can compare the records of employees to determine if there is a pattern of disease in the workplace.
- If done periodically, they can detect the presence of dangerous chemicals in the body *before* actual disease is produced.
- If an employee has developed symptoms—such as headache, dizziness, and fatigue—the tests might point to the cause of the problems.

Before any medical surveillance program is undertaken, the following issues should be considered:

- Medical testing is not a substitute for preventing exposure.
- There should be a scientific and medical basis for doing the tests. The accuracy of the tests should be reasonable.
- The tests should be defined to detect the effects of past and/or present exposures.
- Only the employees at risk from the particular hazard should be tested.
- Employees should be made aware of the limitations of medical testing, regarding both the sensitivity of the tests and the conditions being tested.
- Whoever performs testing and examinations, as well as interpretation of the results, must be a competent physician, preferably board-certified in occupational medicine.
- A full report must be made to each employee about his or her medical examination. With the consent of the employee, this information may be sent to a designated physician or other representatives.
- A report summarizing the results of all testing, without revealing identities of the employees examined, should be available to the employer and representatives of employees.

- The employer should not receive any individual medical results, but rather the physician's opinion on whether evidence of excessive exposure has been detected.
- Employees with conditions that may jeopardize their health in a hazardous situation should be transferred to alternative work without any loss in wages, benefits, or seniority.

Regulations Governing Employee Access to Monitoring and Medical Records

In New Jersey, all private employees (under the federal Occupational Safety and Health Act) and all public employees (under the New Jersey Public Employees Occupational Safety and Health Act [PEOSHA] Access to Medical Records Rule) have the right to obtain a copy of their complete employee medical record and exposure record. Employers must provide this information within 15 days of an employee's request.

Access to medical records is also provided by rules enforced by the New Jersey State Board of Medical Examiners, which licenses physicians in the State of New Jersey. Records available under these regulations include medical records and exposure records. Medical records would include:

- medical histories and questionnaires
- results of laboratory tests
- results of medical exams
- medical opinions, diagnoses, and recommendations
- employee medical complaints
- originals of X-rays plus interpretation
- description of treatments and prescriptions
- records concerning health insurance claims if accessible to the employer by employee name or personal ID number

Exposure records would include the following information:

- industrial hygiene sampling data from personal, area, grab, wipe, or other forms of sampling for:
    - chemicals
    - bacteria, viruses, fungi, etc.
    - noise
    - heat, cold
    - vibration
    - pressure
    - radiation
- results of tests on blood, urine, breath, hair, fingernails, etc. for toxic chemicals

- MSDSs (or, in the absence of MSDSs, any other record which reveals the identity of a toxic substance or harmful physical agent)

For occupational safety and health resources, see Appendix D.

2. INDUSTRIAL HYGIENE

Industrial hygiene is the recognition of environmental factors associated with work and work operations, the evaluation of how these factors affect workers' health and well-being, and the prescription of methods to control or eliminate such factors or minimize their effects.

Physical Forms of Hazardous Substances

The "form" of a substance influences how it enters the body and what damage it can cause. It is important to remember that the form can change during use. The seven main forms of toxic materials are:

1. Solids—Although solid materials are unlikely to be harmful, they can be dangerous if their form changes while they are being used. For example, sanding wood produces wood dust, and plastics can decompose into fumes and gases when heated.
2. Dusts—Dusts are tiny particles of solids that may be inhaled into the lungs. Dust may be created during many work processes (for example, grinding, sanding, and mixing). The most dangerous dust particles are the very small ones, which when inhaled remain in the lungs and cannot be expelled.
3. Fumes—Fumes form whenever a solid material, usually a metal, is heated, volatilizes, then condenses in the air into extremely fine particles. A fume exposure is possible whenever a metal or other solid is melted, poured, cast, burned, welded, or soldered. A solid must be heated above its melting point before it vaporizes; therefore, it is important to know to what temperature a solid is being heated for each process. In most cases, the volatilized solid reacts with oxygen in the air to form an oxide. A common example is zinc oxide fumes from welding on galvanized metal.
4. Liquids—Many hazardous substances, such as acids and solvents, are liquids at room temperature. Many liquids give off vapors which may be inhaled by an employee. Some liquids can damage the exterior surface of the skin, while others are able to pass through the skin.
5. Vapors—Vapor is the gaseous phase of a substance which is a liquid or solid at room temperature. Vapors are usually invisible and may be given off by solids which sublime or liquids which evaporate at room temperature.
6. Mists—Mists are actually tiny droplets of liquids suspended in air and are often visible. A mist can be created by the breaking up of a liquid (by splashing, spraying, foaming, or atomizing).

7. Gases—A gas is a formless fluid which occupies completely any space at room temperature. Some gases are detectable by color or smell, while others can only be detected by special tests.

Chemical Risk

Chemical risk is the probability or likelihood of injury resulting from actual use of the substance in the quantity and manner proposed.

Chemical risk depends on:

• type of toxic effect
• strength of toxic effect
• frequency of exposure (i.e., daily, weekly)
• concentration of exposure (i.e., high concentration or low)
• duration of exposure (i.e., a minute or hours of exposure)
• routes of exposure
• dose actually taken into the body

Routes of Exposure

There are three ways hazardous substances may enter the body:

1. Inhalation—Airborne dust, fumes, vapors, mists, and gases may all be inhaled. They can irritate the skin, eyes, nose, throat, and lungs, or they may also be absorbed into the bloodstream and transported to affect other organs.
2. Skin contact—Skin contact may result in skin damage, sensitization, or irritation. The hazardous substance may also penetrate the skin, enter the bloodstream, and be transported to affect other organs. For example, cleaning products may include corrosive substances that irritate the skin. Other substances, like phenol, may pass into the body through the skin.
3. Ingestion—Exposure via the digestive system is less common and is often overlooked. Food, beverages, cigarettes, and cosmetics can be contaminated with harmful substances. These substances may be ingested when eating, smoking, etc.

A substance may enter the body by more than one route. One substance may even have a variety of effects, depending upon the route by which the body is exposed. Specific organs and parts of the body are more likely to be damaged by specific kinds of hazardous substances. These organs are called target organs for that substance.

Health hazards depend upon dose. This is the total amount of a substance which actually enters the body during exposure. To measure the amount inhaled, air monitoring must take place at the breathing zone, near the nose and mouth, to determine the amount of a substance that can be potentially inhaled. An evaluation of the overall exposure must also consider the potential for skin absorption and ingestion.

Effects of Hazardous Substances

A *local effect* is acquired when a hazardous substance only acts on the part of the body with which it comes in contact. For example, acid burns the skin at the point of contact.

Some hazardous substances do not cause damage at the point where they come in contact with the body, but may be absorbed and carried by the blood to other parts of the body. This is known as a *systemic effect*. An example would be a solvent that enters the body through inhalation, resulting in headaches, dizziness, and nausea.

Once a hazardous substance enters the body, it may have either an acute or a chronic effect, or both. An *acute effect* (such as coughing or irritations of the skin and nose) is one that occurs as an immediate response to exposure. Effects are usually obvious and short-lived; they may be followed by recovery or permanent damage. *Acute exposure* usually occurs as a result of an accident (spill, leak, etc.). *Chronic exposure,* on the other hand, can be more of a problem, as it can exist for years before it is detected. *Chronic effects* (such as lung or kidney disease) are observed after low or repeated exposures, can occur at some time after exposure, and can last for months or years. Consequently, a chronic effect is a delayed (as opposed to immediate) effect of an exposure. Delayed effects are often long-lasting, such as chronic obstructive lung disease resulting from long-term exposure to silica dust. In the case of cancer, limited exposure may result in disease years after the exposure with no previous symptoms.

Latency describes the period of time between exposure to a hazardous substance and the onset of the disease. For example, the latency of lung cancer in workers exposed to asbestos is 20–30 years.

Recognizing, Measuring, and Evaluating Employee Exposure

Use of Senses to Identify a Hazard

Using your senses, including smell, sight, and hearing, is one way to identify a hazard. Be observant. A cloud in the air or wetness may indicate a leak or spill. Stinging eyes, itchy skin, dizziness, or nausea may indicate exposure to a hazardous substance. However, be aware that absence of signs does not mean there is no hazard. For example, carbon monoxide is odorless, colorless, and tasteless, but deadly at high concentrations.

It is sometimes possible to determine that a dangerous amount of a substance is present in the air based on an odor. However, anyone working in an environment that regularly has foul smells knows how quickly one gets used to an odor. For this reason smell is not a reliable way to detect the presence of a hazardous substance. When an employee states, "That odor used to bother me but I don't notice it any more," it probably means that his sense of smell is temporarily impaired or "fatigued." Although the loss of one's ability to smell is usually temporary, it can be permanent. Also, odor is not always a reliable indication of a hazardous condition. Many chemicals are considered toxic at concentrations far below their odor detection levels.

Information About the Process

Knowing the process is one way to identify a hazard. Some exposures are associated with specific processes such as grinding, pouring, dumping, heating, mixing, etc. Knowing exactly what the process entails is an important step to being able to recognize the hazards associated with it.

Labels

To find out information about the health effects of a hazardous substance, it is necessary to know the chemical name and/or the CAS number of the substance. Labels containing an identifier of the hazardous chemical, and hazard warnings, are required by OSHA's Hazard Communication Standard. (See Chapter 2, Section 2.)

Hazardous Substance Fact Sheets

The Fact Sheets distributed by the New Jersey State Department of Health cover important health and safety information for over 2000 substances on the New Jersey Right to Know Hazardous Substance List. The information for each substance includes identification, hazard summary, reason for citation, exposure limits, determining and reducing exposure, medical information, workplace controls, handling and storage, and personal protection equipment. Fact Sheets are available to New Jersey and out-of-state residents and are discussed in Appendix B, Section 2.

Material Safety Data Sheets

MSDSs are health and safety sheets developed by manufacturers for their products. MSDSs, along with Hazardous Substance Fact Sheets, will provide the health and safety information necessary to address a health problem knowledgeably. MSDSs should be requested whenever ordering chemicals. Copies should be kept available and be accessible to employees. (Some MSDSs are incomplete and may not provide enough safety information.)

Industrial Hygiene Sampling Results

A company may use an industrial hygienist to measure levels of specific substances in the work environment. Results of industrial hygiene monitoring are normally maintained at the company facility. This information will help indicate the presence of hazardous substances and should be used to recognize locations in the facility where exposure to hazardous substances occurs.

Industrial Hygiene Monitoring

The purpose of industrial hygiene monitoring is to locate and identify sources of exposures in the workplace so that they can be corrected, and to quantify the exposure of employees to chemicals in the air.

Air monitoring is conducted by industrial hygienists or other persons with specialized training. The hygienist first records relevant data such as process or activity, sources of contamination, and ventilation conditions. Then he or she uses special equipment to measure the levels of substances present in the workplace. Employees should be informed that they have a right to obtain monitoring results under the OSHAct.

Air Samples

Air samples are generally collected in one of three locations:

- at the breathing zone of the worker (personal sample)
- in the general room air (area sample)
- at the operation which is generating the hazardous substance (source sample)

Air samples are collected for two lengths of time. *Grab samples* (instantaneous) measure conditions at one moment in time and can be likened to a still photograph. They give only a picture of conditions at one place at one instant in time. *Continuous samples* (ranging from 20 minutes to 8–10 hours) are taken to evaluate all-day exposure. Continuous samples may be thought of as a motion picture, since they record activity taking place in various places over a period of time. They provide an average of conditions over a period sampled. Short-term (10–15 minutes) samples may also be taken to monitor peak exposures.

Other Sampling Methods

Bulk samples are collected from settled dust in the workplace or from drums or bags of chemicals. Their purpose is to analyze and identify the substances present. For example, bulk samples are used to analyze the percent of asbestos in insulation or dust. A substance constituting greater than one percent of a bulk sample is usually cause for concern.

Wipe samples are used when skin absorption or ingestion is a suspected route of exposure. The purpose is to show whether skin, respirators, clothing, lunch rooms, lockers, etc., are contaminated. It can show which surfaces are clean and which are contaminated. It can also show if some surfaces are more contaminated than others.

Sampling Devices

The general principle of sampling is to collect an amount of a contaminant onto a medium from a known quantity of air. Air samples are collected using a small pump to suck air from the workroom. The pump is attached by tubing to a sampling device which contains the sampling medium (for example, a glass tube containing charcoal). The sampling method used depends on the physical form of the substance:

- Dust—the sampling device is a filter of plastic or paper in a holder.
- Vapors—the sampling device is a glass tube containing activated charcoal and silica gel as a medium.
- Gases—the sampling device is a bubbler containing a fluid medium to dissolve or react with the gas.

The collected samples are sent to a laboratory where the amount of the substance on the sampling medium (filter, tube, etc.) is measured. In some cases, air monitoring is conducted by using a direct reading instrument such as a monitor for carbon monoxide. These instruments can measure the amount of a contaminant without being sent to a laboratory.

Sampling Preparation

In preparing for monitoring, the following questions must be answered:

- *Where* should samples be obtained?
- *Whose* work area should be sampled?
- For *how long* should these samples be taken?
- *How many* samples are needed?
- Over *what period* of work activity should the samples be taken?
- *How* should the samples be obtained?

In answering these questions, the importance of adequate employee input cannot be overemphasized.

Sampling should be done in all areas where employees are exposed to the chemical(s) in question. Sampling devices should be worn by employees representing all job classes, especially those likely to have the worst exposures. This could be on any shift, on the weekend, or during maintenance or shutdown. Only if the worst exposures are measured can one be sure that all employees are protected.

Laboratories and Analytical Methods

A critical step in obtaining accurate sampling results is having samples carefully analyzed by a competent laboratory, using an accurate method. The business of laboratory analysis is highly profitable and many labs have entered the competition. A laboratory which is successfully participating in the NIOSH Proficiency Analytical Testing (PAT) Program should be selected. Furthermore, those participating laboratories can be accredited by the American Industrial Hygiene Association. Note that not all labs are qualified to do every type of analysis. For information on how to find a certified laboratory, see Chapter 5, Section 5.

Selection of an analytical method depends on the type of chemical to be detected. Common methods include:

- dusts—weighing
- metals—atomic absorption
- vapors—gas chromatography

Exposure Limits: Interpretation of Industrial Hygiene Monitoring

This section is based on the booklet "Threshold Limit Values and Biological Exposure Indices" (Cincinnati, OH: American Conference of Governmental Industrial Hygienists, 1988), and on Olishifski, J. B. *Fundamentals of Industrial Hygiene,* 2nd ed. (Chicago, IL: National Safety Council, 1979).

There are two types of exposure limits: Threshold Limit Values (TLVs) and Permissible Exposure Limits (PELs). TLVs are exposure limits established by a nongovernmental group, the American Conference of Governmental Industrial Hygienists (ACGIH). Many of these limits were adopted as legal requirements by OSHA when it started back in 1970. The more recently revised TLVs are often based on the most recent and accurate scientific information. Revised TLVs are almost always lower than the original TLV. (For information on obtaining copies of "Threshold Limit Values and Biological Exposure Indices," see "National Nonprofit Organizations" in Appendix A.)

TLVs represent the levels of chemicals in the air that it is believed most workers can be exposed to day after day without harm. However, some workers may experience discomfort below the TLV, or even possibly develop an occupational illness.

PELs are those employee exposure limits for toxic chemicals published by OSHA as legal standards. A complete list of PELs appears in OSHA's General Standards (CFR 1910.1000, Subpart Z, updated annually). For information on obtaining OSHA standards, see Appendix B, Section 1.

The PELs used by OSHA are based on the TLVs. Both the PEL and TLV lists use two different types of measurement. Concentration of gases and vapors in the air is usually presented as parts per million (ppm). Solids or liquids dispersed in the air, such as mists, dusts, or fumes, are usually expressed as weight per volume: milligrams per cubic meter (mg/m^3). Therefore, it is always important to refer to the column in the PEL or TLV list that has the same units of measurement as your reading.

The idea of "ppm" is not hard to imagine. For example, the PEL for carbon dioxide is 0.5% in air, or 5000 ppm (5000 parts of carbon dioxide per one million parts of air). Levels above the PEL are considered hazardous.

PELs and TLVs are calculated as eight-hour time-weighted averages (TWA). Therefore, occasional levels above the limit are permitted, provided they are balanced by levels below the limit during an eight-hour working day.

For example, if an employee is exposed to ammonia for 4 hours at 40 ppm, 2 hours at 50 ppm, and 2 hours at 60 ppm, the time-weighted average would be:

$$\frac{(4 \times 40) + (2 \times 50) + (2 \times 60)}{8} = \frac{380}{8} = 47.5 \text{ ppm}$$

or below the legal limit (PEL) of 50 ppm of ammonia.

There are limits on how much brief exposures can go above the time-weighted average. The Short-Term Exposure Limit (TLV-STEL) is the maximum level to which workers can be exposed for a period of up to 15 minutes. Exposures at the STEL level should not occur more than four times a day, with at least 60 minutes between consecutive exposures.

Some chemicals have a ceiling TLV (TLV-c), which is the maximum allowable exposure at any one time. When the letter "c" is printed before the exposure limit on the ACGIH TLV list, or before the chemical name on the OSHA PEL list, that exposure limit is the maximum or ceiling value.

For those chemicals which can be absorbed through the skin (about 25% of those on the TLV list), measuring of chemical levels in the air is not sufficient. On the PEL and TLV lists, such chemicals have the word "SKIN" after their name. When a TLV or PEL for a substance is followed by the notation "SKIN," this means that there is evidence that this substance can enter the body by absorption following skin (including eye and mucous membrane) contact. Since a substance's TLV refers only to concentrations in the workroom *air*, this route of entry won't be measured. The "SKIN" notation simply calls attention to the fact that an employee's total exposure to a substance will be increased by skin absorption unless appropriate protective measures are taken.

For some substances, employers have established their own company exposure limits which are lower than the PEL or TLV. For example, the TLV for cutting fluid mists is 5 mg/m^3, but the Ford Motor Company exposure limit is 2.5 mg/m^3.

Important: If the PEL, recommended PEL, and TLV for a chemical are not the same, it is best to use the *lowest* value for the exposure limit.

Exposure limits are intended to protect most employees from health hazards over a working lifetime and are based on the judgment of professionals and scientists who have looked at all the available information on the substance both from animal studies and studies of groups of people who have been exposed.

Exposure Limits: Caveats

Readers should be aware that there are several problems that have been identified with exposure limits:

- For many substances, the information available for choosing the exposure limit is very poor. Since we do not always have solid, reliable information, some guesswork is necessarily involved. Consequently, recommended exposure levels are often changed (and almost always to a lower level) when new information becomes available.
- For some substances, the recommended exposure limit is based only on preventing acute effects. Chronic effects are harder to study. Often, we simply don't have enough information about long-term exposure to know if the substance can cause serious chronic problems at low levels of exposure.
- The way a particular level of a chemical affects one person may be different from the way it affects another person. Some people may be more easily affected than others. So even if the exposure limit protects most people, it may not protect people who are extremely sensitive or more susceptible to a substance.
- TLVs regulate single substances. They do not consider what happens when several chemicals combine to produce effects far more harmful than any one substance produces by itself ("synergistic" effects). Nor do TLVs fully consider what happens when substances are changed in the body to more harmful materials.

3. PERSONAL PROTECTIVE EQUIPMENT

Employees who encounter hazards at the workplace often need to wear personal protective equipment. Depending on the business or industry, they may have to wear safety glasses, face shields and goggles, welding goggles and helmets, hard hats, safety belts and lifelines, safety shoes, special clothing, gloves, earplugs, and respirators.

Personal protective equipment is meant to be used for temporary protection in certain situations, such as cleanups of chemical spills or accident prevention. This equipment can be ineffective and dangerous if it is used improperly or if it is used as a cheap substitute for installing a safe and permanent way of eliminating the hazard from a plant or business.

The workplace you provide for your employees must be designed to be safe. Under OSHA, you are required to provide a hazard-free environment for your employees. Requiring employees to wear personal protective equipment without eliminating the health hazards they face is no solution to the problem. While personal protective equipment is part of the job in some industries—for instance, hard hats to protect construction workers from falling objects—as a rule, it is considered a last-resort type of protection.

When should personal protective equipment be used? A general list includes these conditions:

- emergencies, such as industrial fires, gas leaks, chemical leaks, mine explosions and cave-ins, and special cleanup operations
- accident prevention, such as when a worker puts on a face shield or a respirator just before he or she opens a barrel containing toxic material
- as a means of temporary protection during engineering changes or repairs
- when it is determined, after careful consideration, that engineering controls are not practical

In addition, personal protective equipment should be used as routine safety equipment for some workers. An employee who is exposed to particles, sprays, or splashes needs to wear safety glasses or goggles. Workers who handle toxic or corrosive chemicals and those who work with certain machines and sharp-edged tools must wear gloves. Depending on the industry, special safety shoes are required for certain jobs, such as steel-toe shoes for workers who handle heavy materials. Hard hats are required on construction sites and areas with overhead pipes or beams at head-high levels.

The use of personal protective equipment calls for hazard awareness on the part of the employee. When you train your employees in the safe handling of toxic substances, it is important that they are made aware that the equipment will not eliminate the hazard. Instead, it creates a barrier against hazards. Although some employees may complain that the equipment is uncomfortable or cumbersome, they must not be allowed to modify it or refuse to wear the equipment.

Once you have determined that personal protective equipment is needed for the job, careful attention should be given to how the equipment is selected and how workers are trained to use it. Don't expect to pick up a catalog, order some equipment, and then hand out the equipment to your workers. That is a dangerous and ineffective approach. If you don't already have a policy for the selection, fit, and use of personal protective equipment, you should create one.

The New Jersey State Department of Health recommends considering the following factors before purchasing equipment:

- Make sure that the equipment meets all standards necessary for certification. Among the standards are those written by ANSI on protection for the eyes, face, head, and feet. Standards on respiratory protection are available from NIOSH.
- Ask whether the piece of equipment is capable of protecting a worker against the hazard. While steel-toe shoes can offer protection against excess weight, they don't protect the instep. Thus, metatarsal guards may be necessary. Different gloves are needed for different jobs, and their choice should depend on the chemicals that will be handled. Eye and face protection requirements vary depending on the job.
- Ask your workers for their opinions. Suggest that they field test different brands and models before your company decides which equipment to buy. Comfort and effectiveness are important factors to consider. When workers participate in such decisions, they are more likely to wear the equipment.
- Order the equipment in different sizes so that it will fit all the employees who need to use it.
- Before the equipment is used, make arrangements for regular cleaning, maintenance, and replacement of all the products you are buying.
- Develop a training program to explain to employees the purpose of the equipment, to train them in its use as well as its limitations, and to give them instructions in its care and maintenance.

The following sections will review the most common pieces of equipment that are used to protect workers from the hazards they encounter on their jobs.

Head Protection

Workers who risk being hit by falling objects may be required to wear hard hats or safety helmets. Construction workers, tree trimmers, loggers, electrical utility workers, shipbuilders, miners, and petroleum and chemical workers almost always are required to wear protective headgear.

Safety helmets must be able to protect the worker's head from the shock of a falling object or a bump and also be able to resist penetration. Special helmets are used by electrical workers to protect them from high-voltage shock and burn. Some safety helmets incorporate face and vision protection into the headgear and resemble full-face masks.

The rigid shell of a safety helmet is designed to withstand the force of a blow by using a shock-absorbing band inside the shell. The suspension band and crown straps keep the hat more than an inch above the head. If something were to fall on the wearer, the suspension system would use that space to absorb the shock of the object.

A helmet's headband must be adjusted to the right size. The suspension band must not irritate the worker who is wearing the helmet. Instructions on the proper way to adjust and replace the suspension headband should accompany each helmet. The manufacturer may list warnings on the instructions against using paints or thinners on the shell because such materials could reduce the helmet's ability to protect the wearer.

Each worker should inspect his helmet every day for any damage that might reduce the safety it provides. Any dents or cracks should be noted. This damage and even extremes of temperature that surpass the figures listed in the standards can reduce the degree of safety.

Eye and Face Protection

OSHA requires that employees who work near machines or operations that produce flying particles, sparks, glare, splashes, or harmful radiation wear eye and face protection. When selecting equipment, be sure to choose the right kind of protection for the job. The glasses, goggles, and face shields that OSHA requires include:

- glasses with protective lenses that contain an optical correction for workers who wear eyeglasses off the job. Safety glasses without sideshields are not recommended.
- goggles that can be worn over the worker's own corrective eyeglasses without affecting vision
- goggles that incorporate corrective lenses mounted behind the safety lenses

There are many types of goggles—flexible-fitting goggles, cushioned goggles, plastic eyeshield goggles, and foundrymen's goggles. A number of them are made for specific uses, such as chipping, welding, and cutting. When protection against dust or noncorrosive chemicals is needed, hooded goggles will protect against eye irritation.

If an employee is exposed to splashing from acids, alkalis, molten metals, or sparks, he or she needs protection of the face and neck as well as the eyes. This is best accomplished by using a protective device that combines a face shield and goggles or a helmet with self-contained goggles.

Workers who come into contact with ultraviolet and infrared radiation need goggles with special shading in the lenses to absorb radiation. Those involved with welding, operating a furnace or kiln, brazing, and carbon arcs require goggles with custom-ordered filter shades.

Normal eyeglass frames worn off the job cannot be used to hold safety lenses required for the job. Safety glasses must use specially designed frames.

Eye protection equipment can be kept clean with soap and hot water or a special cleaning solution. When dirt builds up on the lenses, it can strain the eyes. Lenses that are scratched or pitted not only reduce vision, but are more likely to break.

Workers should inspect their eye protectors every day for dirt. They also should note whether the headband is loose, twisted, or worn out. If the elastic has stretched out, the vision protector is likely to slip off the employee's head.

Goggles and safety glasses must be handled with care. It is best to request employees to store the eye protectors in a case when they are not being used.

When eye protection equipment is being reissued to another employee, it must first be cleaned and disinfected. To disinfect a piece of equipment, take it apart and clean all the parts with soap and water. Defective parts should be replaced with new ones. All parts should then be thoroughly wiped with or immersed in a solution of germicidal deodorant fungicide. After soaking the parts for several minutes, remove them, and let each piece dry at room temperature. The parts must not be rinsed because that would cause the germicidal solution to lose its effectiveness.

Protective Clothing

Protective clothing can help shield the body from direct contact with hazardous substances. Heat, cold, chemicals, and dirt are the hazards that are most commonly encountered at work. Some protective clothing covers the entire body, while other items shield only the part of the body exposed to the hazard.

Certain materials are recommended to offer protection against these workplace hazards:

- toxic liquids, including acids and chemicals: rubber, rubberized fabrics, specially coated plastic fabrics
- heat, sparks, molten metal, and welding radiation: leather garments
- radiant heat, including open flames and furnaces: reflective fabrics and insulated fabrics
- cold: insulated fabrics and electrically heated gloves
- dirt, splinters, abrasion: cotton and cotton duck
- dust, toxic dusts: disposable paper suits. (Some disposable suits are made of materials that can be dangerous when they burn. Because their fibers can stick to the skin and cause serious burns, employees who work near open flames or sparks must not wear disposable suits.)

No single fabric is impervious to all chemicals. Constant contact with chemicals or toxic substances breaks down the fabric's ability to protect the worker. Manufacturers of protective clothing can provide data on the degree to which a fabric can repel a certain harmful substance as well as the garment's life expectancy.

Some insulated clothing, including gloves, is made of asbestos fibers that can be released into the air as the garment wears out. When such equipment is needed for the job, it must be monitored carefully for any signs of wear. Most suppliers now offer fireproof or insulating fabrics and gloves that do not contain asbestos.

Proper fit is essential for all protective clothing. If a glove is too large, it can get caught in a machine and pull the employee's hand in with it. Oversized sleeves also can become entangled in machinery. Cuffs on pants or sleeves can retain toxic materials and hold them next to the skin.

In some cases, workers buy their own protective clothing and care for it themselves. But more often, when a business requires its employees to wear protective clothing, management supplies the garments and takes care of cleaning them.

When work clothing is exposed to highly toxic materials, such as lead dust, asbestos, and radioactive substances, the clothing must not be taken home because it will be hazardous to anyone who comes in contact with it. Similarly, clothes containing toxic dusts cannot be sent to commercial laundries. In such cases, providing a plant laundry is the only solution that ensures the safe cleaning of these garments.

Some industries require that workers take off their contaminated clothing in one locker area, take a shower, and put on their street clothing in another locker area. Following these procedures helps prevent the spread of contamination to other workers. Employees who work in settings that call for sterile conditions may have to follow similar procedures to protect the product from outside contaminants.

Good ventilation is necessary in rooms where workers remove their dust-laden clothing. Otherwise, the toxic substance that shakes free from the garments can become airborne and build to a hazardous level in the room's stagnant air.

Unlike the situations listed above, some substances present an immediate danger to the skin. Workers who handle such corrosive chemicals as acids, alkalis, acid gases, and oxidizing agents need to wear clothing that is impervious to liquids, gases, and vapors. Respiratory protection usually must be worn with these types of chemicals. (See Section 4, "Respiratory Protection," below.) Fabrics that are impervious to toxic substances can cause problems if the clothing is not properly ventilated; because perspiration rapidly increases inside the protective clothing, the worker's body temperature can soar to a dangerous level. When clothing incorporates an air line and provides a uniform flow of air to all parts of the protective garment, the worker can remain inside for a longer period of time.

Employees with heart or lung conditions may not be able to safely wear a full-body suit that is hooked up to a respirator. Close medical monitoring is needed for workers who wear clothing made of impervious material.

Foot Protection

Most on-the-job foot injuries are caused by objects that fall from less than four feet, according to OSHA. The agency requires workers who handle heavy materials to wear shoes with safety toes that are impact-resistant. When the area beyond the toes needs protection, the employee should wear instep guards.

Workers in construction who risk puncture wounds from protruding nails need to wear shoes with flexible metal insoles. Shoes with wooden soles are standard safety items for workers who are constantly in contact with hot surfaces as well as those who encounter wet conditions underfoot. People working in the roofing, paving, and hot metals industries wear wooden soles because they do not conduct heat. Wooden-soled shoes also are worn by employees in dairies and breweries.

Lifelines and Safety Belts

When there is the risk of a fall on a job, a personal lifeline system should be used for protection. Workers in the construction industry use lifelines, safety belts, and harnesses because they permit freedom of movement and provide a system of stopping a fall.

To use a lifeline system, the employee wears a belt or harness around the waist. Attached to the harness is a lanyard or rope-grabbing device. If a fall were to occur, the harness would help absorb it by spreading the shock over the shoulders, thighs, and buttocks. The lanyard must not allow the fall to be any deeper than six feet. The rope that is used should have a strength of 5400 pounds and should be composed of a minimum of one-half inch nylon.

Safety belts and harnesses also are used to suspend or retrieve workers from hazardous work areas. Each component of the system being used—including hardware, buckles, hooks, rings, and ropes—must meet specifications. Safety belts must be worn when there is the danger of a fall. Workers who use them should know how to wear them correctly. Belts must be buckled securely and be snug to the body so that the person wearing the belt or harness won't fall out. Lifelines must be inspected regularly for wear and deterioration.

When safety belts and harnesses are not practical, a mesh net should be set up beneath the work area. The net should extend beyond the work area.

When the net is attached to supports, it must be tested for tightness. If the net is too loose, the person falling into it could strike the surface below.

If there is a danger of falling into water while working, the employee should wear a buoyant work vest or a life jacket that is approved by the U.S. Coast Guard.

Ear Protection

An employee who is exposed to a high level of noise on the job is an employee under stress. Prolonged exposure to noise can result in permanent hearing loss. Studies have shown that excessive noise also can cause stress, fatigue, emotional disturbances, and vertigo. The best way to control noise is to build soundproof barriers or booths into the design of the physical plant. A number of engineering remedies also can bring down the noise level. For instance, noise from a vibrating piece of equipment can be lessened by modifying the noisy part. In some cases, the replacement of worn-out gears and bearings or loose belt drives can cut down on noise. Placing sound-absorbent material on the walls and ceilings can lower noise. When such engineering controls are not feasible, hearing protection must be worn.

Under OSHA regulations, exposure to noise during an eight-hour workday cannot exceed 90 decibels as measured on a sound level meter. In some businesses, employees are subject to small spurts of high-intensity noise that rise well above the 90-decibel limit. More often, the exposure is continuous during a shift. Noise levels often surpass the permissible limit for people in certain occupations. Among them are lumber mill workers who hear the constant buzz of saws, metalworkers who labor amidst the sounds of drilling, grinding, and stamping of metal, and printers who work near noisy presses and printing machines.

Noise control through engineering designed to make the workplace quiet is the only way to protect workers. Personal hearing protection can help, but the wearing of earplugs or earmuffs should be viewed only as a temporary solution to the problem of noise on the job.

Earplugs and earmuffs are designed to act as a barrier between the immediate environment and the inner ear. They can reduce sound frequency by a factor ranging from 25 decibels in the low-frequency range to 40 decibels in the high-frequency range. When the sound level is above 120 decibels, no form of ear protection can prevent damage if a worker is continually exposed. OSHA outlaws continuous exposure to noise over 115 decibels.

Various factors can influence the amount of sound that passes through or around the protective device. When there is an air leak between the hearing protector and the skin, the ability to reduce sound is lower. If the material used in the device is of poor quality, it can allow sound to pass through the material itself. The size, design, and composition of the materials in the device all affect the degree to which noise can be reduced.

Some workers prefer wearing earplugs because they are light and easy to carry. However, if an employee has any ear infections or structural abnormalities of the ear canal, he may be unable to wear earplugs. If a worker is sensitive to materials used in the plugs, he may develop an allergic response. Earplugs must be properly fitted by a medical expert if they are to be effective. The size of the plug is determined by the size of the ear canal. Employees often require a different size plug for each ear.

The plugs fit into the ear canal and form a seal that keeps out noise. They are made of soft rubber, plastic, fine glass wool, and wax-impregnated cotton. Plain cotton is ineffective as hearing protection.

Some earplugs are designed to be discarded after one use. Others can last for several months but must be inspected regularly, because they lose their effectiveness over time.

Earmuffs provide good protection against noise because they form a seal over the ear, much like the headphones on a home audio system. The cushioned ear cup that fits over each ear is made of foam rubber or liquid-filled material. Some earmuffs come with a headband and can be worn alone, or they can be mounted on a helmet.

Employees who wear glasses or have beards or sideburns run the risk of reduced protection because their glasses and hair break the seal that keeps out noise. When this is a consideration, special equipment can be ordered.

4. RESPIRATORY PROTECTION

According to federal law, engineering controls must be used to keep the air in the workplace free of contaminants. An employer who has local exhaust ventilation systems in the work areas that need them is in compliance with this requirement. When harmful dusts, vapors, gases, fumes, mists, or sprays enter a worker's breathing zone, it is best to remove them at their source. Reliance on respirators alone is both ineffective and unsafe.

Respirator use is only acceptable if:

- It is a short-term, temporary measure while a plan for engineering controls is being carried out.
- It is needed for infrequently performed procedures for which other precautions have failed.
- It is used as a last resort for a problem that cannot otherwise be solved.
- Respirators are correctly maintained for emergency use.
- The respirator is functioning correctly and provides protection.

Some respirators purify the air a worker breathes by filtering out harmful substances, while others supply breathing air through an air hose or air tank. Many factors must be considered before selecting a respirator, especially the type of contaminant, the estimated amount of contaminant concentration in the work area, and what kind of physical effect the substance has on the human body. Several limitations accompany the use of respirators and they must be fully understood before a respirator is worn. Proper fitting and training in the use of respirators is crucial to maintaining the safety and health of the employee who must wear this equipment. The worker must have full medical clearance before wearing a respirator, because the equipment could cause undue and potentially lethal stress on his or her body.

The improper use of a respirator can result in severe health problems or death. In the event of equipment failure, a respirator wearer must be trained in emergency escape procedures. The standards governing respirator use are strict. As far as the bottom line on costs is concerned, good respirators are expensive to buy and maintain. For all these reasons, it is clear that engineering controls are the best way to protect workers from exposure to toxic substances.

Respirators, however, are not only important, but are required as protection equipment under certain circumstances:

- when work must be done in an area where exhaust ventilation cannot be provided
- when the job to be done is short-term and all other ways of approaching it have been exhausted
- when engineering changes are under way
- when there is an emergency
- when their use can help protect a worker from a potentially dangerous situation (for instance, to guard against breathing of fumes when opening a container of toxic material)

In determining the need for a respirator program at your company, you must know which hazards an employee will encounter in doing the job. Questions to ask include:

- What is the estimated range of concentration of toxic material being produced in the air in the work area?
- Is only one contaminant being emitted into the air, or is there a number of toxic substances and by-products? Can you identify every chemical and material that enters the air in the work area? (Respirators must not be used when unknown chemicals are present or the chemical concentration in the air is uncertain.)
- How much time must be spent doing the job? How much physical activity is required while wearing a cumbersome respirator?
- Is the concentration of contaminant immediately dangerous to life or health? (If so, exposure can either seriously injure or kill a person within a few minutes, or it can enter the body and eventually cause cancer.)
- Is there a deficiency of oxygen in the work area?

- Is there a way to warn a respirator wearer that the chemical contaminant has begun to pass through the canister supplying his air? Does the contaminant itself have any warning properties (e.g., odor, taste, and eye irritation) to alert the wearer that danger is imminent?
- Can the contaminant be absorbed through the skin? If it can, the job will require additional protective equipment.

All of the above questions need to be asked before selecting respirators for employees.

The selection of equipment must be based on the kind of hazard that is present. Thus, a respirator that is designed to filter dust will not protect a worker who is exposed to vapors, fumes, and gases.

Types of Respirators

Two kinds of respirators are available. *Air-purifying* respirators contain filtering media to prevent the worker from inhaling minute particles of solid airborne contaminants such as sawdust, coal dust, or beryllium oxide. Included in this group are respirators that filter out certain liquid droplets, such as ammonia and chlorine. *Air-supplying* respirators supply air or oxygen to the worker through an air hose or a self-contained tank that is carried by the employee.

Air-Purifying Respirators

Particle-removing air-purifying respirators are made of a facepiece that contains a mechanical filter. The filter may be composed of cotton, wool, synthetic fibers, glass or mineral fibers, or a combination of all these materials. A filter whose fibers are more tightly packed is more efficient, but it also can result in increased resistance to breathing.

This kind of respirator is useful only when it is known that there is adequate oxygen in the work area and that the particles it is filtering out are the only harmful substances in the room. Proper design of the respirator and proper fit are important. If air is allowed to leak around the filter, or if the respirator does not make a perfect seal against the employee's face, its ability to protect the employee is reduced. These respirators must be inspected often and kept clean.

These respirators come in half-face mask and full-face mask varieties. A disposable single-use face mask now is available for some purposes. Also available is a battery-powered filter respirator.

Half masks offer no protection to the eyes and they have a greater likelihood of leaking than a full-face respirator. They are attached to the face by means of elastic straps.

Full masks must form a seal around the employee's face and consequently cannot be worn if the worker has a beard. An employee who wears corrective eyeglasses would also have an air leak if he wore his glasses with the respirator. This problem can be solved by purchasing a mask that allows a set of corrective lenses to be mounted into the facepiece of the respirator. These masks attach to the face by means of a head harness with adjustable straps.

As a worker breathes, valves inside the respirator send outside air into the filter and into the nose, and then direct the exhaled air outside the mask. The inhalation valve opens when it pulls in air, but closes when the worker exhales. The exhalation valve must close quickly in order to keep toxic material out of the respirator. Valves must be checked regularly for signs of leaks.

Gas- and vapor-removing air-purifying respirators offer protection against gases or vapors by reacting with the harmful substance in the outside air and removing it from the air inhaled by the worker. This type of equipment comes with a cartridge or canister that is filled with a sorbent, or substance that pulls in and then absorbs, adsorbs, or reacts with the contaminant in the atmosphere.

The sorbent in a respirator may be made to protect against only one kind of compound, or it may be able to work against several substances, such as acid gases or organic vapors. Employees must check a canister before beginning a job in order to make sure that the sorbent offers protection against the harmful substance that will be encountered. (For a full list of sorbents and the materials with which they should be used, refer to the *Respiratory Protective Devices Manual*, published by the American Industrial Hygiene Association and American Conference of Governmental Industrial Hygienists.)

The manufacturer's instructions on each canister regarding shelf life must be followed carefully, because the chemical that makes the product work becomes less effective once the expiration date has passed. Also, a sorbent can deteriorate after a canister has been opened and used for even a short time, because it tends to absorb moisture and other substances in the air. Canisters that are outdated and those that have been opened for a period of time after even a brief usage should be discarded.

Employees should follow the manufacturer's instructions on how long the canister can be used in a harmful atmosphere before it loses its effectiveness. The only warning that the sorbent is depleted is the odor of gas in the canister. Some canisters change color when the sorbent is nearly run out.

These air-purifying respirators will not work if the oxygen supply is inadequate in the work area, nor will they offer protection against hazardous solid particles in the air. However, masks that combine a filter for harmful particles and a sorbent for vapors, gases, ammonia, and the oxidation of carbon monoxide are available. This kind of mask is a universal gas mask. This mask will not offer protection in an area where there is no oxygen.

Half-face masks are available for quick jobs in areas where the concentration of gas or vapor is low. Full-face masks are usually required because the airborne contaminant often is harmful and irritating to the eyes as well as to the respiratory system.

Air-Supplying Respirators

When oxygen is low or absent from the atmosphere and a worker must enter the area, the only permissible equipment that can be used is a supplied-air respirator. This protective equipment feeds air or oxygen from an outside source through an air line to the worker's mask or helmet.

The supply of compressed air flows from a cylinder of air or oxygen or from a compressor that takes in air from outside or from the air supply for the building. Compressors that are water-lubricated provide the best source of compressed air. When compressors driven by other sources are used, there is a risk of carbon monoxide entering the breathing air system. For instance, a compressor that is driven by a gas or diesel engine must be checked to make sure that exhaust gases from the engine are directed away from the air pump's intake valve.

Atmosphere-supplying respirators employ several facepieces, depending on the need. Half- and full-face masks, helmets that cover the head, hoods, air-ventilated blouses, and full-body suits all can provide an air supply to the worker.

When an air hose is used to supply air, the supply of fresh air can be delivered a few different ways. A face mask hooked up to an air blower that is hand-operated by another employee can provide an air supply and also an observer who can rescue the worker if an accident occurs. For less hazardous conditions, a hose mask without an air blower or an air line respirator attached to a cylinder of compressed air can be used.

Air line respirators employ two modes of operation: continuous flow and demand. The continuous flow mode provides a regular flow of air into the mask. If there is any leakage, it will be excess air from the mask entering the atmosphere. The demand mode uses a valve to regulate the flow of air into the mask. It draws in air from the cylinder only when the wearer inhales. The signal to send air comes as a result of the negative pressure produced inside the mask with each inhalation. When the demand method is used, the mask must fit perfectly against the face to prevent the inhalation of contaminated air. The demand system usually has a bypass switch that allows the wearer to change over to continuous flow.

Three types of self-contained breathing apparatus are available for conditions that require a worker to carry his own supply of air as he moves around an area. There is (1) a demand and pressure-demand type, (2) an apparatus that recirculates the air to conserve the supply of oxygen, and (3) a type using a chemical source of oxygen that becomes available to the wearer when it absorbs moisture and carbon dioxide during exhalation.

Approved Respirators

When a respirator is approved for use, it has a label that indicates its approval by the Mine Safety and Health Administration (MSHA) and NIOSH. Tests are conducted on each respirator and the minimum performance standards are set. After a respirator gains approval, it is added to the list of approved devices.

OSHA specifies the use of approved respirators if they are available. An approved respirator contains an identification number on each unit and a label that identifies the hazard it protects against and notes the limitations of the respirator. The components that are approved for use with the respirator also are indicated on the approval label.

All of the respirators that NIOSH has certified are listed in NIOSH Publication 80-144, "NIOSH Certified Equipment List." The American National Standards Institute also lists guidelines in its "American National Standard Practices for Respiratory Protection Z88.2-1980."

Selection and Fitting of Respirators

The points that must be considered when selecting a respirator were addressed earlier in this section. Getting a proper fit for each employee is essential. The face seal must be tight, but it should not leave deep indentations on the face or make turning the head difficult.

There are two kinds of tests that can determine whether a respirator fits properly—qualitative tests and quantitative tests. The following is a summary of fit tests recommended by the National Safety Council in its publication *Protecting Workers' Lives.*

Qualitative Fit Tests

Qualitative tests include the positive pressure test, the negative pressure test, and the odor test. They are simple and easy to perform, and indicate whether a mask is comfortable for the worker who must wear it.

The *positive pressure test* entails removing the exhalation valve cover when possible, blocking the valve, and gently exhaling into the facepiece. If a slight positive pressure is built without any outward leakage of air from the face seal, then the fit is considered to be satisfactory. After the test, the exhalation valve cover must be carefully replaced.

The *negative pressure test* calls for blocking the inlet opening of the canister by covering it with the palm. Then inhale gently so that the facepiece collapses slightly, and hold the breath for about 10 seconds. If the facepiece remains slightly collapsed, the fit of the respirator is considered satisfactory.

The *odor test* consists of passing an irritant such as banana oil (isoamyl acetate) three inches from the respirator seal. If the wearer is able to smell the oil, his or her respirator is leaking. Testers should know that irritant smoke tubes can be used only with particulate cartridges.

Quantitative Fit Tests

Quantitative tests, on the other hand, are considered more accurate for checking the fit of a respirator. In these tests, the employee enters a closed booth wearing a respirator with probes. A test solution is then sent inside the booth and a special device measures the amount of solution in the booth and in the respirator. A comparison of the two levels of concentration indicates whether there is any leakage. OSHA's standard for lead requires quantitative testing of respirators.

Both methods of fit testing complement each other. Employees get the best protection when both testing methods are used.

Training Employees in Respirator Use

Up-to-date training is the basis of an effective respirator program. Such a program should emphasize the ongoing training of the employee who is in charge of the program. This person must be constantly aware of any changes in substances that are used in the workplace, whether newly approved equipment would better protect employees using respirators, and what kind of needs are being expressed by the workers who wear this equipment.

A respirator program has many important facets: the selection and fitting of respirators, training in how they are used, instructions on their care and maintenance, and safety sessions that emphasize their protective abilities as well as their limitations.

It is recommended that supervisors and foremen take part in respirator fitting and training programs, even if they rarely use respirators. This provides a good example for employees and demonstrates that management strongly supports the safe use of respirators.

Before a worker uses a respirator, he/she needs careful instruction on how the unit works, an explanation of the parts it contains, practice in properly attaching the device to his/her head, and a test run of the respirator to ensure that it is the proper fit. (See "Selection and Fitting of Respirators," above.) The trainer also must show the employee how to inspect the respirator before each use. The importance of inspecting the equipment before using it also must be conveyed by the trainer.

Employees don't like to wear respirators. They are cumbersome and limit a worker's ability to move. Some workers are afraid of respirators. Managers must have a high level of interest in a respirator program if it is to be successful. It is important to provide working conditions that do not require respirators. When respiratory protection is needed under short-term conditions, workers are more likely to accept the wearing of a respirator to do the job.

Maintenance of Respirators

Regular inspection, cleaning, and maintenance is necessary whenever respirators are used. They must be checked for signs of wear and deterioration. Every part of the respirator must be scrutinized to make sure that it is working properly. Valves, filters, hoses, facepieces, and headstraps all require regular inspection. Respirators that are used for emergencies and self-contained breathing apparatus need to be inspected on a regular monthly schedule because they must be ready for use under extremely dangerous conditions.

An employee who uses a small respirator can inspect his or her own equipment. His or her training in the use of a respirator for the job must emphasize the need for regular inspection. While larger plants may have a special department for the cleaning and maintenance of respirators, small plants often do not. In this case, a worker can clean and replace any worn-out parts under the guidance of a safety engineer or industrial hygienist.

Large pieces of respiratory protective equipment are much more complex and must be inspected by a safety engineer. A trained cleaning and maintenance expert must dismantle, wash, dry, and replace parts when necessary, install new filters or canisters, and reassemble each respirator on a regular basis.

After a respirator has been cleaned and is designated to be stored for a while, it should be placed in a plastic bag to keep out dust. New units should be kept in their original packing until they are ready to be used. Employees who carry their respirators with them must be taught how to properly store these units. By stuffing a respirator into a cramped space, an employee can ruin the face seal he or she needs to protect the air supply.

Respirator Limitations

Working under hot conditions can cause discomfort and fatigue to an employee wearing a respirator. This problem can be relieved by wearing a full-body suit that supplies air to all areas of the body. One compact device that supplies cool air can be worn on the belt or on a harness.

Fatigue quickly leads to reduced efficiency. Factors that contribute to fatigue include increased resistance to breathing, reduced vision caused by the facepiece, and anxiety over being inside a respirator. Fatigue can be reduced by giving the worker frequent breaks.

The ability to communicate is limited when a worker wears a respirator. If a worker tries to talk, the movement of his/her jaw can break the seal and allow contaminated air to enter the breathing zone.

Vision is limited when a respirator is worn. Half- and full-face masks obstruct downward vision, making it easy for a worker to trip or fall. Full-face masks virtually cut out peripheral vision. Workers who require prescription glasses find that the protective seal of the mask to the face is broken by the temple piece. Because of this, they need to be fitted with special glass kits for full-face masks. Wearing glasses with a half-face mask also can be a problem, because the glasses can slide around above the mask or they can enter the area between the face and the mask, thus breaking the protective seal.

Workers with heart or lung conditions should not wear respirators. The weight of a respirator could be too much strain for an employee with a heart problem. The increased breathing resistance that comes with wearing a respirator could be dangerous to an employee with a respiratory problem such as emphysema. When an employee's ability to use this equipment is questioned, a medical examination must be performed by a physician.

OSHA Minimal Acceptable Respirator Program

If respirators are used in your plant, the following guidelines for an acceptable respiratory protection program are required by OSHA:

1. Written standard operating procedures governing the selection and use of respirators shall be established.
2. Respirators shall be selected on the basis of hazards to which the worker is exposed.
3. The user shall be instructed and trained in the proper use of respirators and their limitations.
4. Where practicable, the respirators should be assigned to individual workers for their exclusive use.
5. Respirators shall be regularly cleaned and disinfected. Those issued for the exclusive use of one worker should be cleaned after each day's use, or more often if necessary. Those used by more than one worker shall be thoroughly cleaned and disinfected after each use.
6. Respirators shall be stored in a convenient, clean, and sanitary location.
7. Respirators used routinely shall be inspected during cleaning. Worn or deteriorated parts shall be replaced. Respirators for emergency use, such as self-contained devices, shall be thoroughly inspected at least once a month and after each use.
8. Appropriate surveillance of work area conditions and degree of employee exposure or stress shall be maintained.
9. There shall be regular inspection and evaluation to determine the continued effectiveness of the program.
10. Persons should not be assigned to tasks requiring use of respirators unless it has been determined that they are physically able to perform the work and use the equipment. The local physician shall determine what health and physical conditions are pertinent. The respirator user's medical status should be reviewed periodically (for instance, annually).
11. Approved or accepted respirators shall be used when they are available. The respirator furnished shall provide adequate respiratory protection

against the particular hazard for which it is designed in accordance with standards established by competent authorities. (All respirators must have a seal of approval from NIOSH/MSHA.)

References

Levy, B. S. and D. Wegman, Eds. *Occupational Health: Recognizing and Preventing Work-Related Disease* (Boston: Little, Brown and Company, 1983).

Protecting Workers' Lives: A Safety and Health Guide for Unions (Chicago: National Safety Council, 1983).

"New Jersey Worker and Community Right to Know Act, Education and Training Program Guide" (New Jersey State Department of Health, 1986).

The Industrial Environment—Its Evaluation and Control (Washington, DC: U.S. Department of Health and Human Services/Public Health Service, National Institute for Occupational Safety and Health, 1973).

"Personal Protective Equipment" (U.S. Department of Labor, Occupational Safety and Health Administration, 1985).

5. EMPLOYEE EDUCATION AND TRAINING

In order for employees to work in a safe and efficient manner, it is necessary to inform them of proper equipment use, work procedures, and policies. This process occurs through formal and informal employee education and training programs. Workers are required by state Right to Know laws and OSHA to be formally educated and/or trained in the following areas:

- some state Right to Know laws (public employees only in New Jersey)
- OSHA Hazard Communication Standard (1910.1200)
- OSHA Comprehensive Standards
- OSHA Respiratory Protection Standard (1910.134)
- OSHA Hazardous Waste Operations and Emergency Response Standards (1910.120)

For training requirements under state Right to Know laws, contact your state Right to Know program. To find out what OSHA standards pertain to your industry, call your OSHA area office (see Appendix B) or consult *Code of Federal Regulations* Title 29.

The educational resources available to employers are abundant. Although small industries usually do not have the personnel or funds to conduct sophisticated training programs, the ample supply of low-cost or free resources should be adequate to conduct a quality program.

Education/training sessions can be developed and conducted by staff trainers or outside consultants. Teaching aids useful in program sessions include literature, audiovisuals, equipment demonstrations, and discussions. To assist employers in designing their own educational programs, there are guide-lines available for developing programs on the Right to Know law, OSHA Standards, and hazard communication. (See Appendix D, Section 1.)

According to the *OSHA Handbook for Small Businesses,* training is required for all current employees and new employees when they start work. Each employee needs to know the following:

- No employee should undertake a job until he or she has received instructions on how to do it safely and has been authorized to perform that job.
- No employee should undertake a job that appears to be unsafe.
- Mechanical safeguards (machine guards) must be kept in place.
- Each employee should report all unsafe conditions encountered during work.
- Any injury or illness suffered by an employee, even a slight one, must be reported to management at once.
- Any safety rules that are a condition of employment, such as the use of safety shoes or eye protection, should be explained clearly and then enforced.

Indicators which might show a need for training or retraining are:

- excessive waste or scrap
- high labor turnover
- an increase in the number of "near misses" which could have resulted in accidents
- a recent upswing in actual accidents experienced
- high injury and illness incidence
- expansion of business and/or new employees
- a change in process, or introducing a new process

The following guidelines provide suggestions for developing a quality training program (and reducing the threat of potential lawsuits and other liabilities):

1. *Define training requirements:* Determine what training employees need. Identify all areas where workers who are potentially exposed to hazardous materials/waste need to be educated.

2. *Assign a training coordinator:* A person at the plant should be in charge of coordinating, developing, and reviewing training programs. In small industries, this duty may be performed by an environmental manager. This job entails the development of a library of pertinent materials, and keeping a filing system for worker training records.

3. *Use a combination of trainers:* The best facility training programs are those which use both in-house trainers and third-party trainers. Some in-house conflicts of interest may exist with using only company trainers, while external trainers can bring fresh ideas and

perspective to a training program. In-house trainers are valuable to programs because of their familiarity with site-specific hazards and problems.

4. *Know your trainer:* Many unqualified people offer training programs. In order to ensure the trainer is competent:

- Review credentials (e.g., resume).
- Check with prior clients.
- Ask about affiliations and organizations.
- Check out the trainer's track record; ask other trainers.
- Talk with the trainer directly to obtain a general idea on information presented, and the perspective of the course.
- Find out the specific price for specific services.

5. *Module approach:* The module approach to training is optimal because: (1) small groups of workers are trained; (2) employees receive a specific and digestible amount of information; (3) it is employee-oriented and not "generic"; and (4) it provides employees with building blocks of information.

6. *Provide the time:* In order to ensure worker safety and reduce the company's liabilities, it is important not to cut corners on time or quality.

7. *Retraining/review is a must:* Ongoing training programs are necessary. These sessions may include lecture reviews, hands-on training, simulations, and advanced training sessions. When possible, use post-testing and certification exams to ensure a minimum level of competence among workers.

8. *Obtain corporate involvement:* A superior training program needs corporate backing. Trainers, resources, equipment, and lost work time cost money. Management sets the tone for health and safety.

9. *Extend network:* Evaluate and assess the training of your contractors. Untrained or incompetently trained transporters can make mistakes with your company's hazardous waste.

10. *Trainers do not have all the answers:* Make sure trainers are accountable. Many now include a disclaimer stating that they are responsible for the presentation of material, but not for the ultimate actions or behavior of workers on the job.

Reprinted from the *Environmental Manager's Compliance Advisor*, Issue No. 172, Dec. 2, 1985, with permission of the publishers (Business and Legal Reports, 64 Wall Street, Madison, CT 06443).

For resources in employee education and training, see Appendix D, Section 1.

6. SAFETY AND HEALTH COMMITTEES

The following material was adapted and updated from *Organizing a Safety Committee* (OSHA 2231, 1975).

Since the total responsibility for a company's safety and health program is often too much for one individual to handle, many industrial establishments have developed safety and health committees to detect unsafe plant conditions and health hazards. In nonunion establishments, these are organized by management. In unionized workplaces, safety and health committees are usually set up as joint union-management committees.

Committee Structure

What kind of committee do we need? How many members should be on it? These are among the first questions that are considered when management decides to tackle plant hazards through an active safety and health committee. The answer most frequently given by experts is to let the size of the plant and its hazard potential(s) dictate the type and size of the committee.

Small companies often prefer to use one central safety and health committee, with the manager as chairperson and his or her key employees as members. For example, a small (80 employees) Salem, Oregon company that produces batteries finds one committee enough to do the job. Management and employee representatives together regularly inspect every operation in the building, to make the workplace safe and healthful.

The committee chairperson should be a person whose authority exceeds the authority of each member of the group. This gives a fair guarantee of (1) effective, controlled action to follow committee findings, and (2) access to the next higher level of management via the committee chairperson.

Unionized Workplaces

In unionized workplaces, it is most effective to involve the union (as a representative of the employees) in safety and health work. This will involve having the union designate employee representatives to joint labor-management safety and health committees and having management choose management representatives. Ideally, there should be an equal number from both groups. The union may ask that this arrangement, as well as the roles and responsibilities of the joint committee, be formally written down in the collective bargaining agreement (contract). The local union may also have its own union safety committee, which would include the union members of the joint committee and/or other people.

On joint committees, it is also often effective to have the role of the chairperson alternate between labor and management representatives.

Role of the Safety and Health Committee

To be successful, a management-worker safety and health committee should be involved in the initial planning of the safety and health program and should be involved in the operation of the program. Definite policies should be established at the time the committee is organized. The following functions should be considered:

- establishing a process for handling suggestions and recommendations of the committee
- inspecting a selected area of the workplace each month for the purpose of detecting hazards
- conducting regularly scheduled meetings to discuss accident and illness prevention methods, safety and health promotion, hazards noted on inspections, injury and illness records, and other pertinent subjects
- investigating accidents and near-accidents as a basis for recommending means to prevent recurrence
- providing information on safe and healthful working practices to managers, supervisors, and other employees
- recommending changes or additions to improve protective clothing or equipment
- developing or revising rules to comply with current safety and health standards

Guidelines

In helping the safety and health committee to work most effectively, the following guidelines may be useful:

- Keep the safety and health committee small so that every member can participate actively.

- If a job looks too big for a small group, subdivide it, allocating portions of the tasks to small committees. Everyone on the committee should be actively involved.

- Committee members don't approach the task knowing all the answers. However, they have gained valuable experience and knowledge about shop floor conditions which they bring to the committee's work.

- Management should provide the committee with its direction, its goals, and its limits. In a plant in Jacksonville, Florida, for example, the manager's safety and health committee is charged with the review and approval of all proposed new installations before they are put into operation. A supervisory staff member chairs the committee; supervisors and hourly work representatives are members. The committee encourages all employees to report safety and health hazards as they see them and recommend better practices. While reviewing the comments of both employees and its own members, this properly directed committee establishes goals and limits.

- To keep members actively involved, the committee should meet at least once a month and carry out assignments between meetings, encouraging meaningful research and observation of plant safety and health hazards.

- In advance of each meeting, the secretary should notify each member. A good practice is to combine the meeting reminder with the delivery of the minutes of the last meeting.

- Meetings should not last for more than an hour. In order to accomplish brief meetings, the secretary should prepare a tight agenda with the chairperson, who conducts the meeting in the order presented in the agenda.

- A good meeting program should be established. First the meeting should be called to order by the chairperson. Next, revisions should be made of minutes from the previous meeting, followed by a signing of the attendance sheet and reports on past assignments. Suggestions and discussion of work that needs to be done come next. At all times, the members should be stimulated to come forward with ideas and suggestions. Before the meeting is adjourned, specific tasks should be assigned and accepted with deadline dates for completion noted in the minutes.

It has been a common pitfall for committees to stray from their specific territory. It takes diplomacy to keep spirited discussions on course. There must be a firm rule that no issues other than those concerning safety or health may be brought up at meetings.

At times it is the chairperson who must be kept on course. It is easy to slip out of the leader's role as coordinator and try to do the legwork of members. When committee officials step out of character, they frustrate the members, and the committee ceases to function under its original purpose.

All members should be encouraged to participate and to receive recognition for their work. Many companies have a system of sending copies of minutes to employees to keep them posted on the efforts and achievements of the committee.

Avoid showpiece committees. Beware of committees lacking authority. The best advice to members of a committee lacking management support is to disband. Committees with no real power degenerate into self-perpetuating institutions with all of the attendant abuses. Lame-duck committees tend to talk about problems rather than solve them.

The committee should also not avoid responsibility. This is a major claim of weakness held against committees. Give new members responsibilities; the sooner they become involved, the faster they will develop the skills required to be effective members of the committee. Don't let a handful of employees perpetuate themselves in committee duty.

The key to a successful committee is to plan a careful system of rotation. One common rotation method is use of overlapping terms, so that there is always a certain percentage of experienced members on the committee.

7. PROMOTING EMPLOYEE HEALTH

The four leading causes of death in adults in the United States are heart disease, cancer, accidents, and stroke. In addition to the suffering caused, and the financial burden placed on families, the loss to industry as a result of disability or death due to disease or accidents is substantial.

Although the causes of disease are varied and complex, there is increasing evidence that a person's lifestyle has a significant effect on one's health. Factors such as smoking, alcohol consumption, diet, and exercise all contribute to a person's risk of disease. The need to educate people on the steps necessary to prevent disease is now well accepted by health care providers.

Health care costs in the United States are constantly rising. In the past decade, costs have increased an average of 13% per year. This burden weighs heavily on employers, as they are the largest private purchasers of health care.

Illnesses and injuries often result from heart conditions, respiratory disease, obesity, stress, unsafe work conditions, hypertension, alcoholism, or drug dependency. These illnesses and their accompanying costs are reflected in:

- absenteeism
- tardiness
- sick pay
- accidents on and off the job
- reduced productivity
- workers' compensation insurance
- disability insurance
- early pension payments

- loss of personnel through early death
- employee replacement costs

Actual improvement in the health status of employees is one very direct way of lowering business costs of worker illness and injury. It is well documented that healthful lifestyles and a safe work environment can prevent disability, sickness, and premature death.

In the past few years, employers have shown an increasing interest in maintaining and improving the health of their employees. This has been largely due to the belief that healthy employees are more productive and experience less absenteeism than unhealthy workers. Encouraging employees to stay well has taken form in worksite wellness programs, which have the potential for addressing a variety of health issues.

The costs involved in setting up a health promotion program can frequently be offset by reduced employee medical expenses. To measure the cost-effectiveness of a health promotion program, businesses have begun to measure program start-up and operating costs against the dollar benefits of the program. There is evidence that worksite programs can save dollars while also providing other less measurable benefits to employers and employees resulting from improved mental and physical well-being.

Implementing a Wellness Program

Developing a wellness program requires several steps to ensure success. First, it is important to officially establish and communicate to employees the company's commitment to health promotion. It is equally important to involve employees in the actual planning and design of the company's program, so that the program can meet the needs of employees, as well as motivate employees to become involved in the program. This can be done via a health planning committee of management, union representatives, and other interested employees.

The committee's task should be to determine the types of health promotion programs which suit both employees' health needs and the company's resources. A needs assessment can be done via a survey of employee health interests and problems. Other means of assessing needs include analysis of type and frequency of illness as reflected in health insurance, disability, and workers' compensation claims; analysis of worksite accident records, workplace conditions, and hazardous materials exposures which may contribute to illness or injury; or reviewing national and regional statistical trends to identify the probable health risks for employees (which requires compiling data on age, sex, ethnic background, and other indicators for the employee population).

While it is more feasible for large businesses to conduct comprehensive wellness programs, many small business owners and managers also recognize the value of providing health promotion activities for their employees. Due to a variety of resources available to the employer, implementing health promotion activities can be done with minimal cost and effort. (To assess employee illness costs, see "Illness Cost Audit: The Overall Picture," below.)

A variety of topics can be covered in an employee wellness program. An employer may decide to offer employees a selection of activities and topics, or choose to focus on a specific area of concern. The most common areas for health promotion activities include weight control, nutrition, stress management, physical fitness, smoking cessation, and alcohol and drug counseling. Other topics include safety practices, seatbelt use, CPR, healthy back information, women's health issues, and disease prevention. Programs can take the form of lectures or classes, health screenings, exercise activities, videocassette presentations, or demonstrations.

Management can also support wellness activities by implementing steps that encourage healthy behaviors in employees. For example, some small businesses create policies such as bonuses for employees who participate in health promotion activities, flex-time for employees, or designated non-smoking areas. (See "Smoking Legislation in New Jersey," Chapter 2, Section 5.) Informal practices such as inviting families to attend wellness programs, supplying healthy foods in vending machines, and adding extra time onto lunch hour for exercise also encourage healthy behaviors. A supportive physical environment may take form in removing cigarette machines from the facility; placement of bike racks, showers, and lockers to encourage exercise; or displaying signs and posters to reinforce health concepts in common areas such as the lunchroom.

Since small businesses have limited resources, there are certain factors that should be considered when planning activities. According to "Wellness in Small Businesses," a report developed for the Washington Business Group on Health, small businesses that have implemented successful programs usually have these program factors in common:

- small in scope
- simple
- inexpensive
- family-oriented
- part of a "healthy" company

There are diverse possibilities for wellness activities in the worksite. Following are some examples of activities that have been used by various small businesses:

- videocassette presentations on health topics during lunch hour
- lunchtime speakers to discuss various health topics

- health screenings by local agencies for high blood pressure, diabetes, cancer, lung ailments, glaucoma, etc.
- exercise programs such as lunchtime exercise classes, walking clubs, discounted memberships for employees at local health clubs, reserving a local pool for use by employees after work hours, counseling on physical fitness assessments and fitness goals, and/or rewards, such as T-shirts with company logo, for meeting exercise goals
- smoking cessation/reduction programs offered one to two times a year, removal of cigarette machines from workplace, and/or rewards or bonuses for employees who quit smoking
- replacing snack machines containing "junk foods" with healthy snacks such as fruit and decaffeinated coffee, and/or serving healthy foods in cafeterias

Finally, keep in mind employees may be suspicious of or may not participate in health promotion programs if the pressure for change is placed only on them and not on the work environment, too. Efforts at reducing physical stressors (toxic chemicals, noise) or psychological stressors (poor communication, lack of employee input) can improve health as well as convince employees the company is serious about health promotion.

Illness Cost Audit: The Overall Picture

The following material is reprinted with permission from "Wellness in the Workplace—A Strategic Approach for Employers" (Golden Empire Health Systems Agency, Sacramento, CA, 1983).

To determine whether or not to invest in health promotion, a company should first examine its current illness costs. An illness cost audit can provide information on past and current health-related costs, as well as establish baseline data for the future evaluation of a health promotion program.

Using an illness cost audit form, companies can be provided with an insight into the real costs of employee illness and injury (Figure 4.1). Information is gathered on a per-employee basis. It is important to look at business health costs on a per-employee basis so that the data are not distorted by growth or reduction in a company's work force. Thus, a company can use this audit to actually compare its current costs with previous year's costs. This type of comparison can provide the information upon which to project cost trends into the future.

Since each company has its own set of unique circumstances, not all line items may be applicable to every company (see instructions). Also, some companies may have special programs and benefits that are not included in this audit. (Lines 8 and 9 of the audit have been reserved for that purpose.)

An audit provides a company with a rough estimate of expenditures to identify *illness cost trends*. If a company's total illness cost per employee as

	Cost 5 Years Ago	Cost Last Year	Cost This Year	Projected Cost in 5 Years
1. Average cost of health insurance premium per employee				
2. Average cost of sick pay per employee				
3. Average cost of substitute workers (due to absenteeism) per employee				
4. Average cost of disability insurance per employee (not SDI)				
5. Average cost of workers' compensation insurance per employee				
6. Cost of early pension payments distributed per employee				
7. Health-related replacement costs distributed per employee				
8. _____				
9. _____				
10. Total Illness Cost per Employee				
11. Total Company Illness Costs				

Figure 4.1.   Illness Cost Audit form.

indicated in line 10 of the audit is rising, a health promotion program in the worksite can be a worthwhile investment.

Since wellness is a long-term effect, it is important to remember that many improvements that may result from a health promotion program may not become obvious for a number of years. In addition, monetary savings will not reflect the less tangible effects of a health promotion program, such as increased job productivity or improved job satisfaction.

Illness Cost Audit Instructions

The following directions should be used to complete the Illness Cost Audit (Figure 4.1):

1. *Average cost of health insurance premium per employee:* Determine the employer's insurance premium cost for the identified year, and divide that total by the number of employees covered by the insurance plan for that same year.

   This figure would need to be adjusted if employee benefits, deductibles, or copayments have changed from one year to the next. This adjustment is important because a constant per-employee cost over a several-year period does not necessarily reflect stabilized costs. More than likely it may reflect reduced employee benefits, or increased employee copayments or deductibles.

   A company may also want to adjust the cost if a rise in premiums is *directly* attributable to large claims unrelated to the overall health status of the employee (e.g., automobile accident claims). If relevant, this type of adjustment should also be applied to the final figure in items 2, 3, 4, 6, and 7 below.

2. *Average cost of sick pay per employee:* Add up the company's total sick pay expenditures for the identified year, and divide by the total number of workers eligible for sick pay during that same year. Since the monetary value of sick pay is equivalent to salary costs, the figures should be adjusted to account for increases due solely to rising wages.

3. *Average cost of substitute workers (due to absenteeism) per employee:* Add up the company's total substitute worker expenditures due to absenteeism for the identified year, and divide that total by the number of workers employed by the company. Similar to item 2, the figures should be adjusted to account for increases due solely to rising wages.

4. *Average cost of disability insurance per employee:* (Since State Disability Insurance [SDI] is an employee expense, it is not in-

cluded as part of this business cost audit.) Some companies have disability insurance plans as part of their employee benefit package. This line item is designed for companies which have these benefits.

Determine the employer's disability insurance premium cost for the identified year, and divide that total by the number of employees covered by the disability plan for that same year. This figure would need to be adjusted if employee disability benefits have changed from one year to the next. This adjustment is important, because a constant per-employee cost over a several-year period does not necessarily reflect stabilized costs. It may reflect reduced employee disability benefits, or increased employee copayments or deductibles.

5. *Average cost of workers' compensation insurance per employee:* Add up the company's workers' compensation insurance costs for the identified year, and divide that total by the number of employees covered by workers' compensation for that same year. A company's workers' compensation insurance rates are periodically evaluated and are increased if the company's use of the compensation fund is rising. To be useful, the final figure should be adjusted to account for the insurance increases due to expanded benefits and rising wages rather than due to the increased use of the compensation system.

6. *Cost of early pension payments (for health reasons) distributed per employee:* Pension plans vary considerably from company to company. Under some pension plans, early retirement costs the company more money over the long run than retirement at the usual age. Under other pension plans, this is not the case. This item should be completed *only* if an early (illness-related) retirement results in higher overall costs to the company. Under that circumstance, determine the employer costs of early pension payments for the identified year and divide that total by the number of employees covered under the company's pension plan for that same year.

7. *Health-related replacement costs distributed per employee:* These expenses refer to the costs associated with permanently replacing a disabled or deceased employee. For each employee replaced due to health reasons, determine the approximate recruitment, training, and lost productivity costs associated with his or her departure and replacement. Divide that total by the number of workers employed by the company.

8. and 9. *Direct or indirect illness costs not identified in this audit should be itemized here.*

10. *Total Illness Cost Per Employee:* Add items 1 through 9 to determine the approximate *illness cost per employee* for each identified year. Since these figures are given on a per-employee basis, useful comparisons can be made among current year costs, past costs, and projected future costs.

11. *Total Company Illness Costs:* To determine the total company illness cost, multiply the figure in line 10 by the number of permanent employees (part-time and full-time) in the company's work force. This cost figure can provide a company with a rough estimate of the total dollars spent on illness for the identified year. If the size of the company's work force remains constant over the years, the annual figures in this item can be compared with one another to detect illness cost trends over time.

References

Behrens, R. "Wellness in Small Businesses" (Washington Business Group on Health, 1985).
Behrens, R. "Worksite Health Promotion: Some Questions and Answers to Help You Get Started" (U.S. Department of Health and Human Services, Office of Disease Prevention and Health Promotion, 1983).
"Wellness in the Workplace—A Strategic Approach for Employers" (Golden Empire Health Systems Agency, 1983).
"Lifestyle Improvement for Employees Program" (New Jersey State Department of Health, 1984).

8. SANITATION

This section was adapted with permission from *Protecting Workers' Lives: A Safety and Health Guide for Unions* (Chicago: National Safety Council, 1983).

Design and Procedures

Good workplace sanitation is important not only for the reduction of health hazards, but to help maintain or increase employee morale. The following outline suggests proper sanitation design and procedures:

- Receptacles for garbage should be provided in all workplaces, provided with tight-fitting covers, and cleaned on a regular basis.

- Workplaces should be constructed in such a way as to prevent entrance or harborage of rodents, insects, or vermin. If infestation is detected, extermination should be done.
- Proper cleaning of workplaces and equipment should be conducted.

Water

Potable (drinkable) water must be available in all workplaces, for washing, cooking, drinking, washing of food and food preparation equipment, and personnel service rooms. Ice should also be made up of potable water and be sanitary. Open containers (e.g., barrels, pails, or tanks) from which water must be poured or dipped are prohibited, as are common drinking cups. Where disposable cups are provided, a sanitary receptacle for disposal is required. Nonpotable water for firefighting or industrial use should be clearly marked as such.

Toilets

All workplaces must have toilet facilities, their number based on the number of employees of each sex. Where toilet facilities can be occupied by only one person, separate toilets for men and women are not required. Each toilet room must provide toilet paper with holders and with covered receptacles supplied for facilities used by women. Sewage disposal shall not interfere with the sanitary condition of the toilet facilities.

Washing Facilities

According to OSHA requirements (29 CFR 1910.14(d)), for each three toilet facilities, one lavatory must be provided. Kept in sanitary condition, each lavatory must provide hot and cold running water with soap or other cleansing materials provided. Hand towels, warm air blowers, or continuous cloth toweling with proper receptacles for disposal must also be provided. When showers are required, these standards must be followed:

- Soap or other cleansing materials should be provided.
- Individual clean towels should be provided.
- One shower for each 10 employees of any one sex must be available.
- Hot and cold running water must flow into one line (shower head).

Change Rooms

When employees are required to wear protective clothing, change rooms, with storage facilities for street clothes, are to be provided. Protective clothing is to be stored in separate facilities from street clothes.

CHAPTER 5

Sources of Information and Assistance

1. HEALTH AND SAFETY CORE REFERENCE LIBRARY

The following publications are basic elements of a health and safety reference library, and can be obtained at very little cost. The total cost of this first list is $52. (Prices are subject to change.)

Pocket Guide to Chemical Hazards. NIOSH-OSHA, Washington, DC, 1987. DHEW (NIOSH) Publication No. 78-210. Superintendent of Documents, U.S. Government Printing Office, Washington, DC 20402-9325. $7
"Threshold Limit Values and Biological Exposure Indices." ACGIH. Cincinnati, 1988–89. American Conference of Governmental Industrial Hygienists, 6500 Glenway Avenue, Cincinnati, OH 45211, (513) 661-7881. $4.50
"Fourth Annual Report on Carcinogens." Research Triangle Park, 1985. NTIS No. PB85-134633. Available from National Toxicology Program, Public Information Office, MD B2-04, Box 12233, Research Triangle Park, NC 27709. Free publication.
NIOSH Publications. Complimentary single copies are available from the National Institute of Occupational Safety and Health, 4676 Columbia Parkway, Cincinnati, OH 45226. Ask to be put on their mailing list.
Emergency Response Guidebook. Washington, DC, 1987. Training Unit DHM-51, Federal, State and Private Sector, Initiatives Division, Office of Hazardous Materials Transportation, Research and Special Programs Administration, U.S. Department of Transportation, Washington, DC 20590. Publication No. DOT P 5800, single copy available upon request.

"Occupational Lung Disease—An Introduction." American Lung Association. Free publication available from your local American Lung Association Chapter. (See Appendix D, Section 2.)

Work is Dangerous to Your Health. Stellman, J. and S. Daum. New York, 1973. Random House Inc., 400 Hahn Road, Westminster, MD 21157. $5.95 plus $1 shipping and handling.

Ventilation: A Practical Guide. Clark, N., T. Cutter, and J. McGrane. Center for Occupational Hazards, 1984. Available through NYCOSH, 275 7th Avenue, 25th floor, New York, NY 10001. $7.50 plus 15% postage and handling.

Office Work Can Be Dangerous to Your Health. Stellman, J. and M. Heniffin. Pantheon Books, 1983. Available through Women's Occupational Health Resource Center, 117 St. John's Place, Brooklyn, NY 11217 (718) 230-8822. $6.95

"A Worker's Guide to the New Jersey Workers' Compensation Law." 1984. Rutgers Labor Education Center, IMLR, Ryders Lane, New Brunswick, NJ 08903, (201) 932-9242. $2.50

Noise Control: A Guide for Workers and Employers. 1984. American Society of Safety Engineers, 1800 E. Oakton Street, Des Plaines, IL 60018. $15 non-members plus $2 shipping.

These other publications are excellent reference books which you can purchase to add more depth to your library:

The Condensed Chemical Dictionary. Hawley, G. G. New York, 1987, 11th ed. Van Nostrand Reinhold Company, Inc., 115 5th Ave., 4th Floor, New York, NY 10003. $52.95

Guidelines for the Selection of Chemical Protective Clothing. ACGIH. Cincinnati, 1983. American Conference of Governmental Industrial Hygienists, 6500 Glenway Avenue, Cincinnati, OH 45211. $35

Industrial Ventilation. A Manual of Recommended Practice. 19th ed. ACGIH. Cincinnati, 1986. American Conference of Governmental Industrial Hygienists, 6500 Glenway Avenue, Cincinnati, OH 45211. $20

Basic Industrial Hygiene, A Training Manual. Brief, R. S. Akron, 1975. American Industrial Hygiene Association, P.O. Box 70477-T, Cleveland, OH 44190. $25 plus $1.50 shipping and handling.

Emergency Handling of Hazardous Materials in Surface Transportation. Student, P. J. Washington, DC, 1987. Association of American Railroads, Bureau of Explosives, Publications, 50 F Street NW, Washington, DC 20001. $25

Fire Protection Guide on Hazardous Materials. 1986. Available from National Fire Protection Association, Batterymarch Park, Quincy, MA 02269, (800) 344-3555. $44 plus $2.85 handling.

Safety Management: Accident Cost and Control, 4th ed. Grimaldi, J., and R. Simmonds. Homewood, IL: 1988. Richard D. Irwin, 1818 Ridge Rd., Homewood, IL 60430. $39.95 plus $3 shipping and handling.

The Industrial Environment—Its Evaluation and Control. 1973. Department of Health and Human Services. Available from Superintendent of Documents, U.S. Government Printing Office, Washington, DC 20402-9325. Publication No. 017-001-00396-4, $26

The following publications provide comprehensive technical information on specific environmental and occupational health topics. These books are available through Lewis Publishers, Inc., P.O. Drawer 519, 121 S. Main Street, Chelsea, MI 48118, (800) 525-7894.

Lowrys' Handbook of Right-to-Know and Emergency Planning (1988). Lowry, G. G. and R. C. Lowry. $74.95

Quantitative Risk Assessment for Environmental and Occupational Health (1986). Hallenbeck, W. H. and K. M. Cunningham. $39.95

Toxicology: A Primer on Toxicology Principles and Applications (1988). Kamrin, M. A. $27.50

Low-Level Radioactive Waste Regulation—Science, Politics and Fear (1988). Burns, M. E. $39

Methods of Air Sampling and Analysis (1988). Lodge, J. P., Jr. $80

Indoor Air and Human Health (1985). Gammage, R. B. and S. V. Kaye. $44.95

Principles of Hazardous Materials Management (1988). Griffin, R. D. $45

Air Quality (1985). Godish, T. $39

Handbook of Hazardous Waste Management for Small Quantity Generators (1988). Phifer, R. W. and W. R. McTigue, Jr. $39.95

Toxic Air Pollution (1987). Lioy, P. J. and J. M. Daisey. $49.95

The following journals also serve as core reference materials in a health and safety library:

Applied Industrial Hygiene is published monthly by Applied Industrial Hygiene, Inc., a subsidiary of the American Conference of Governmental Industrial Hygienists, Inc., 6500 Glenway Ave., Bldg. D-7, Cincinnati, OH 45211-4438, (513) 661-7881. For 12 issues, subscription rate is $60 for nonmembers.

American Industrial Hygiene Association Journal is published monthly for $75 per year for nonmembers. It is available through the American Industrial Hygiene Association, 475 Wolf Ledges Parkway, Akron, OH 44311-1087, (216) 762-7294.

American Journal of Public Health, published monthly, is the official journal of the American Public Health Association, Inc., 1015 15th St., NW, Washington, DC 20005, (202) 789-5600. The publication is available for $60 per year.

JAPCA, The International Journal of Air Pollution Control and Waste Management is published monthly by the Air Pollution Control Association at P.O. Box 2861, Pittsburgh, PA 15230, (412) 323-3444. Subscription rate is $170 per year.

2. OSHA ONSITE CONSULTATION FOR THE EMPLOYER

Adapted from "Consultation Services for the Employer," OSHA 3047.

Employers who want help in recognizing and correcting safety and health hazards and in improving their safety and health programs can receive it free from a consultation service largely funded by the Occupational Safety and Health Administration. In New Jersey, this service is delivered by the state government through the Department of Labor.

Besides helping employers to identify and correct specific hazards, consultants provide guidance in establishing or improving an effective safety and health program and offer training and education for the employer, the employer's supervisors, and employees. The service is provided chiefly at the worksite, but limited services may be provided away from the worksite.

Primarily targeted for the smaller businesses in industries with higher hazards or with especially hazardous operations, the safety and health consultation program is completely separate from the inspection effort. In addition, no citations are issued or penalties proposed, and the service is confidential.

Benefits

If a manager can identify the hazards in the workplace and is aware of appropriate resources, he or she will be in a better position to comply with job safety and health requirements. The more you know about the hazards in your workplace—and ways to improve them—the better you can comply with job safety and health requirements. The consultation program provides free professional advice and assistance without the need to hire additional staff. When a consultant helps set up or strengthen a workplace safety and health program, safety and health activities become routine considerations rather than crisis-oriented responses. In addition, if you have a complete examination of your workplace, correct all identified hazards, post notice of their corrections, and institute the core elements of an effective safety and health program, your company may be excluded from general schedule OSHA enforcement inspections for one year.

Procedure

Consultants carefully study each workplace and the employer's safety and health program in order to apply their professional expertise to its specific problems and unique operations.

Comprehensive consultation services include the following:

- an appraisal of all mechanical and environmental hazards and physical work practices
- an appraisal of the present job safety and health program or the establishment of one
- a conference with management on findings
- a written report of recommendations and agreements
- training, and assistance with implementing recommendations
- follow-up to assure that any required corrections are made

Getting Started

Consultation starts with the employer's request, by telephone call, correspondence, or contact during a promotional visit conducted by a consultant. Some services, such as a review of proposed new production processes from a safety and health point of view, may be conducted at locations away from the employer's worksite. When onsite services are requested, the consultant will confer with the employer at the outset regarding the specific needs or concerns described. The consultant may also research any special problems mentioned in the initial contact before scheduling a visit to the establishment. At your request, assistance may also include education and training for you, your supervisor(s), and your employees.

Opening Conference

Upon arrival at the worksite for a scheduled visit, the consultant will briefly review his or her role during the visit and may, if requested, review with the manager the company's safety and health program. Employee participation is encouraged during the consultation process.

Walking Through the Workplace

During this process, the manager and the consultant will examine conditions in the workplace. The consultant will identify any specific hazards and provide advice and assistance in establishing or improving the safety and health program and in correcting any hazardous conditions identified.

The consultant will either study the entire operation or focus on those areas, conditions, or hazards for which assistance has been requested. He or she will also offer advice and assistance on other safety or health hazards that might not be covered by current federal or state OSHA standards, but that still pose safety or health risks to employees.

In a complete review of a company's operation, the consultant will look for mechanical and physical hazards by examining the structural condition of the building, the condition of the floors and stairs, and the exits and fire protection equipment. During the tour of the workplace, he or she will review the layout, checking for adequate space in aisles and between machines, check equipment such as forklifts, and examine storage conditions. Control of electrical hazards and machine guards will also be considered.

The consultant will check the controls used to limit worker exposure to environmental hazards such as toxic substances and corrosives, especially air contaminants. He or she will check to see if all necessary technical and personal protective equipment is available and functioning properly. Also, the consultant will note any problems workers may encounter from exposure to noise, vibrations, extreme temperatures, or unusual lighting on approaches, and offer means and techniques commonly used for the elimination or control of hazards.

Work practices, including the use, care, and maintenance of hand tools and portable power tools, as well as general housekeeping, are of interest to the consultant. He or she will want to talk with the manager and with workers about items such as job training, supervision, safety and health orientation and procedures, and the maintenance and repair of equipment.

In addition, the consultant will want to know about any ongoing safety and health program the firm has developed. If the firm does not have a program or would like to make improvements, the consultant will, at the company's request, offer advice and technical assistance on establishing a program or improving it. Management and worker attitude toward safety and health will be considered in this analysis, as well as current injury and illness data. The consultant will need to know how management and employees communicate about safety and health as well as any in-plant safety and health inspection programs.

Closing Conference

Following the walkthrough, the consultant will meet with the manager in a closing conference. This session offers the consultant an opportunity to discuss measures that are already effective and any practices that warrant improvement. During this time, the manager and the consultant can discuss problems, possible solutions, and time frames for eliminating or controlling any serious hazards identified during the walkthrough.

The consultant may also offer suggestions for establishing, modifying, or adding to the company's safety and health program in order to make such programs more effective. Such suggestions could include worker training, changing work practices, methods for holding supervisors and employees accountable for safety and health, and various methods of promoting safety and health.

Hazard Correction and Program Assistance

After the closing conference, the consultant will send the manager a written report explaining the findings and confirming any correction periods agreed upon. The report may also include suggested means or approaches for eliminating or controlling hazards, as well as recommendations for making the company's safety and health program effective. Consultants can be contacted for additional assistance at any time.

Inspection Exemption

Employers who receive a comprehensive consultation visit, correct all identified hazards, and institute the core elements of an effective safety and health program may be awarded a certificate of recognition by OSHA signifying a one-year exemption from general schedule enforcement inspections. However, inspections prompted by an employee complaint or by a fatality or catastrophe would not be exempted under this program.

Summary

The Consultation Program provides several benefits for an employer. On-site consultants will:

- help identify hazards in the workplace
- suggest approaches or options for solving a safety or health problem
- identify sources of help available to the employer if any further assistance is needed
- provide a written report that summarizes these findings
- assist in developing or maintaining an effective safety and health program
- offer training and education for employers and employees at the workplace, and in some cases away from the site
- under specified circumstances, recommend a company for recognition by OSHA and a one-year exclusion from general schedule enforcement inspections

Consultants *will not:*

- issue citations or propose penalties for violations of federal or state OSHA standards
- routinely report possible violations to OSHA enforcement staff
- guarantee that any workplace will "pass" a federal or state OSHA inspection

Consultation is a highly successful program that generates employer response. To receive additional information, contact the consultation program available in your state.

OSHA/State Consultation Project Directory

Alabama
Onsite Consultation Program
P.O. Box 6005
University, Alabama 35486
(205) 348-3033

Alaska
Division of Consultation & Training
  LS&S/OSH
Alaska Department of Labor
3301 Eagle Street
Pouch 7-022
Anchorage, Alaska 99510
(907) 264-2599

Arizona
Consultation and Training
Division of Occupational Safety &
  Health
Industrial Commission of Arizona
P.O. Box 19070
800 West Washington
Phoenix, Arizona 85007
(602) 255-5795

Arkansas
OSHA Consultation
Arkansas Department of Labor
1022 High Street
Little Rock, Arkansas 72202
(501) 682-4522

California
CAL/OSHA Consultation Service
525 Golden Gate Avenue, 2nd Floor
San Francisco, California 94102
(415) 557-2870

Colorado
Occupational Safety & Health
  Section
Institute of Rural Environmental
  Health
Colorado State University
110 Veterinary Science Building
Fort Collins, Colorado 80523
(303) 491-6151

Connecticut
Division of Occupational Safety &
  Health
Connecticut Department of Labor
200 Folly Brook Boulevard
Wethersfield, Connecticut 06109
(203) 566-4550

Delaware
Occupational Safety and Health
Division of Industrial Affairs
Delaware Department of Labor
820 North French Street, 6th Floor
Wilmington, Delaware 19801
(302) 571-3908

District of Columbia
Office of Occupational Safety &
  Health
DC Department of Employment
  Services
950 Upshur Street, NW
Washington, DC 20011
(202) 576-6339

Florida
Onsite Consultation Program
Bureau of Industrial Safety and
  Health
Department of Labor & Employment
  Security
LaFayette Building, Room 204
2551 Executive Center Circle, West
Tallahassee, Florida 32301
(904) 488-3044

Georgia
Onsite Consultation Program
Georgia Institute of Technology
O'Keefe Building, Room 23
Atlanta, Georgia 30232
(404) 894-3806

Guam
OSHA Onsite Consultation
Government of Guam
3rd Floor, International Trade Center
P.O. Box 9970
Tamuning, Guam 96911
(671) 646-9246

Hawaii
Division of Occupational Safety &
  Health
830 Punchbowl Street
Honolulu, Hawaii 96813
(808) 548-7510

Idaho
Safety & Health Consultation
  Program
Boise State University
Department of Community &
  Environmental Health
1910 University Drive, MG110
Boise, Idaho 83725
(208) 385-3283

Illinois
Division of Industrial Services
Department of Commerce &
  Community Affairs
100 West Randolph Street, Suite 3-
  400
Chicago, Illinois 60601
(312) 917-2339

Indiana
Bureau of Safety, Education &
  Training
Indiana Division of Labor
1013 State Office Building
Indianapolis, Indiana 46204-2287
(317) 232-2688

Iowa
Consultation Program
Iowa Bureau of Labor
1000 East Grand Avenue
Des Moines, Iowa 50319
(515) 281-5352

Kansas
Kansas Consultation Program
Kansas Department of Human
  Resources
512 West 6th Street
Topeka, Kansas 66603
(913) 296-4086

Kentucky
Consultation and Training
Kentucky OSH Program
Kentucky Labor Cabinet
U.S. Highway 127, South, Bay 4
Frankfort, Kentucky 40601
(502) 564-6895

Louisiana
Consultation Program
Louisiana Department of Labor
1001 North 23rd Street
P.O. Box 94094
Baton Rouge, Louisiana 70804-9094
(504) 925-6005

Maine
Division of Industrial Labor
Maine Department of Labor
Labor Station 45
State Office Building
Augusta, Maine 04333
(207) 289-3331

Maryland
Consultation Services
Division of Labor & Industry
501 Saint Paul Place
Baltimore, Maryland 21202
(301) 333-4218

Massachusetts
Consultation Program
Division of Industrial Safety
MA Department of Labor &
   Industries
100 Cambridge Street
Boston, Massachusetts 02202
(617) 727-3567

Michigan (Health)
Special Programs Section
Division of Occupational Health

Michigan Department of Public
   Health
3500 N. Logan, P.O. Box 30035
Lansing, Michigan 48909
(517) 335-8250

Michigan (Safety)
Bureau of Safety and Regulation
Michigan Department of Labor
7150 Harris Drive, P.O. Box 30015
Lansing, Michigan 48909
(517) 322-1814

Minnesota (Safety)
Consultation Division
Department of Labor & Industry
444 Lafayette Road, 5th Floor
St. Paul, Minnesota 55101
(612) 297-2393

Minnesota (Health)
Consultation Unit
Department of Public Health
717 Delaware, S.E.
Minneapolis, Minnesota 55440
(612) 623-5100

Mississippi
Onsite Consultation Program
Division of Occupational Safety &
   Health
Mississippi State Board of Health
307 West Lorenz Blvd.
Jackson, Mississippi 39213
(601) 987-3981

Missouri
Onsite Consultation Program
Division of Labor Standards
Department of Labor & Industrial
   Relations
621 East McCarthy Street
Jefferson City, Missouri 65101
(314) 751-3403

Montana
Montana Bureau of Safety & Health
Division of Workers' Compensation
5 South Last Chance Gulch
Helena, Montana 59601
(406) 444-6424

Nebraska
Division of Safety, Labor & Safety
  Standards
Nebraska Department of Labor
State Office Building
301 Centennial Mall, South
Lincoln, Nebraska 68509-5024
(402) 471-4717

Nevada
Training and Consultation
Division of Occupational Safety &
  Health
4600 Kietzke Lane, Building D-139
Reno, Nevada 89502
(702) 789-0546

New Hampshire
Onsite Consultation Program
New Hampshire Department of
  Labor
19 Pillsbury Street
Concord, New Hampshire 03301
(603) 271-3170

New Jersey
Division of Workplace Standards
N.J. Department of Labor
Office of Occupational Health &
  Safety Consultation
CN 953
Trenton, NJ 08625
(609) 984-3507

New Mexico
OSHA Consultation

Occupational Health & Safety
  Bureau
1190 St. Francis Drive, Room 2200
Santa Fe, New Mexico 87504-0968
(505) 827-2885

New York
Division of Safety and Health
New York State Department of
  Labor
One Main Street
Brooklyn, New York 11201
(718) 797-7646

North Carolina
North Carolina Consultative
  Services
North Carolina Department of Labor
Shore Memorial Building
214 West Jones Street
Raleigh, North Carolina 27603
(919) 733-2360

North Dakota
Division of Environmental
  Engineering
ND State Department of Health
1200 Missouri Avenue, Room 304
Bismarck, North Dakota 58502-
  5520
(701) 224-2348

Ohio
Division of Onsite Consultation
Ohio Department of Industrial
  Relations
P.O. Box 825
2323 West 5th Avenue
Columbus, Ohio 43216
(614) 481-5697

Oklahoma
OSHA Division
Oklahoma Department of Labor

1315 Broadway Place, Room 301
Oklahoma City, Oklahoma 73103
(405) 235-0530 ext. 240

Oregon
Consultation Program
Employer Services, Room 204
Labor and Industries Building
Salem, Oregon 97310
(503) 378-2890

Pennsylvania
Indiana University of Pennsylvania
Safety Sciences Department
Uhler Hall
Indiana, Pennsylvania 15705
(412) 357-2561/2396
(800) 382-1241 (toll-free in state)

Puerto Rico
Occupational Safety & Health Office
PR Department of Labor & Human
    Resources
505 Munoz Rivera Avenue, 21st
    Floor
Hato Rey, Puerto Rico 00918
(809) 754-2134/2171

Rhode Island
Division of Occupational Health
RI Department of Health
206 Cannon Building
75 Davis Street
Providence, Rhode Island 02908
(401) 277-2438

South Carolina
Onsite Consultation Program
Consultation & Monitoring, SC
    DOL
3600 Forest Drive
P.O. Box 11329

Columbia, South Carolina 29211
(803) 734-9599

South Dakota
S.T.A.T.E. Engineering Extension
Onsite Technical Division
South Dakota State University
Box 2218
Brookings, South Dakota 57007
(605) 688-4101

Tennessee
OSHA Consultative Services
Tennessee Department of Labor
501 Union Building, 6th Floor
Nashville, Tennessee 37219
(615) 741-2793

Texas
Occupational Safety & Health
    Division
Texas Department of Health
1100 West 49th Street
Austin, Texas 78756
(512) 458-7287

Utah
Utah Safety & Health Consultation
    Service
P.O. Box 45580
Salt Lake City, Utah 84145-0580
(801) 530-6868

Vermont
Division of Occupational Safety &
    Health
Vermont Department of Labor &
    Industry
120 State Street
Montpelier, Vermont 05602
(802) 828-2765

Virginia
VA Department of Labor & Industry
P.O. Box 12064
205 N. 4th Street
Richmond, Virginia 23241
(804) 786-5875

Virgin Islands
Division of Occupational Safety &
   Health
VI Department of Labor
Lagoon Street
Frederiksted, Virgin Islands 00840
(809) 772-1315

Washington
Voluntary Services
WA Department of Labor &
   Industries
1011 Plum Street, HC-462
Olympia, Washington 98504
(206) 586-0961

West Virginia
West Virginia Department of Labor
State Capitol, Building 3, Room 319
1800 E. Washington Street

Charleston, West Virginia 25305
(304) 348-7890

Wisconsin (Health)
Section of Occupational Health
WI Department of Health & Social
   Services
1414 E. Washington Avenue, Room
   112
P.O. Box 309
Madison, Wisconsin 53701
(608) 266-8579

Wisconsin (Safety)
Division of Safety and Buildings
Wisconsin Department of Industry,
Labor and Human Relations
1570 East Moreland Boulevard
Waukesha, Wisconsin 53186
(414) 521-5063

Wyoming
Occupational Health and Safety
State of Wyoming
604 East 25th Street
Cheyenne, Wyoming 82002
(307) 777-7786

3. OSHA VOLUNTARY PROTECTION PROGRAM

The purpose of the Voluntary Protection Program (VPP), supported by the Occupational Safety and Health Administration, is to emphasize the importance of, encourage the improvement of, and recognize excellence in employer-provided, site-specific occupational safety and health programs. These programs are comprised of management systems for preventing or controlling occupational hazards. The systems not only ensure that OSHA's standards are met, but go beyond the standards to provide the best feasible protection at a site.

When employers apply and achieve approval for participation in the VPP, they are removed from programmed inspection lists. This frees OSHA's inspection resources for visits to establishments that are less likely to meet the requirements of the OSHA standards. VPP participants enter into a new relationship with OSHA in which safety and health problems can be approached cooperatively when and if they arise.

Participation in the VPP does not diminish existing employer and employee rights and responsibilities under the OSHAct. In particular, OSHA does not intend to increase the liability of any party at an approved VPP site. Employees or any representatives of employees taking part in an OSHA-approved VPP program are not assuming the employer's statutory or common law responsibilities for providing safe and healthful workplaces or undertaking in any way to guarantee a safe and healthful work environment. The programs included in the VPP are voluntary in the sense that no employer is required to participate but that any employer may volunteer for application to one of the VPPs. Compliance with OSHA standards and applicable laws remains mandatory.

Program Description

The VPPs are voluntary programs which provide recognition, and removal from programmed inspection lists, to qualified employers. They emphasize the importance of worksite safety and health programs in meeting the goal of the act "to assure so far as possible every working man and woman in the Nation safe and healthful working conditions . . . " through official recognition of excellent safety and health programs, assistance to employers in the efforts to reach a level of excellence, and the use of the cooperative approach to resolve safety and health problems.

The VPPs consist of two major programs, "Star" and "Try," plus a demonstration program to permit demonstration and/or testing of experimental approaches which differ from the two established programs. In addition, within the Star and Try programs there are some variations between general industry and construction industry requirements.

By approving an applicant for participation in the VPP, OSHA recognizes that the applicant is providing, at a minimum, the basic elements of ongoing systematic protection of workers at the site, making routine federal enforcement efforts unnecessary. The symbols of this recognition are certificates of approval and the right to use flags showing the program in which the site is participating. The participant may also choose to use the program logos in such items as letterhead or award items for employee contests.

Note that while OSHA will remove approved worksites from programmed inspection lists, the worksites are not exempt from valid, formal employee safety and health complaint inspections, investigations of significant chemical spills/leaks, or fatality/catastrophe investigations. OSHA will provide the opportunity to work cooperatively with the agency both in the resolution of safety and health problems and in the promotion of effective safety and health programs through such means as presentations before organizations (for example, the National Safety Congress). Each approved site will have a designated OSHA contact person to handle information and assistance requests.

Eligible Applicants

1. *Site management:* Management at a site which is either independent or part of a corporation can make application to the VPP for that site.

2. *Corporate management:* The management of a corporation may apply to the VPP on behalf of one or more sites in the corporation. This type of application is particularly appropriate when one or more aspects of the site safety and health program are provided by corporate staff.

3. *General contractors and organizations providing overall management at multi-employer sites:* At multi-employer sites, such as in the construction industry, the only eligible applicant is the one which can control safety and health conditions of all employees at the site, such as the general contractor or owner.

4. *Organizations representing groups of small businesses in the same industry* (at the three- or four-digit SIC level in a limited geographical area): All sites must meet the requirements and will be subject to onsite review.

Applications for all VPPs must be accompanied by certain assurances describing what the applicant will do if the application is approved for participation in one of the VPPs.

Program Options

Star

Open to any industry, Star is designed for companies in high-hazard industries with comprehensive, successful safety and/or health programs. Companies that are in the forefront of employee protection, as indicated by three-year average injury incidence and lost workday case rates at or below the national average for their industry, may participate. The approach may be management initiative or employee participation. Because of the special hazards at construction worksites, construction firms are limited to the employee participation alternative. Star participants are evaluated onsite every three years, although their incidence rates are reviewed annually.

Try

Try may be either a stepping-stone to Star or a means of testing alternate safety and health strategies which might ultimately be included in a Star program. As in Star, participants may cover safety and/or health and may use an employee participation approach or a management initiative approach. While there are no fixed rate requirements for Try, applicants must show specific goals for reducing rates to levels below the average for their industry. Try participants are evaluated annually.

Praise

Firms with good safety records in low-hazard industries are eligible for the Praise program. Covering safety only, the program is designed to recognize and encourage excellent, effective programs in industries not targeted for OSHA inspections. To participate, a company must have a five-year average lost workday case and total injury incidence rate at or below that of its specific industry. Since this is a recognition program, there is no evaluation, but incident rates are reviewed annually.

Program Responsibilities

1. *Application review:* Each applicant undergoes a review of its safety and/or health program, including an onsite examination of its safety and health records, a review of its inspection history, if any, and an assessment of site conditions. For Star and Try applicants, OSHA also conducts interviews of management and employees, if appropriate. The onsite portion of the review generally requires about a day and a half.

2. *Evaluation:* Annual evaluations for Try participants and three-year evaluations for Star companies compare injury and/or illness rates to industry rates, measure the satisfaction of all groups at the site, and assure that the companies continue to meet the requirements.

3. *Contact person:* For each participant, OSHA provides a contact person to provide assistance and expedite variance applications.

4. *Inspections:* OSHA retains responsibility for inspections in response to formal, valid employee complaints and workplace fatalities and catastrophes.

4. NIOSH HEALTH HAZARD EVALUATION PROGRAM

One of the activities of NIOSH involves Health Hazard Evaluations (HHE), or on-the-job investigations of reported worker exposures to toxic or potentially toxic substances. Made as a direct response to requests by management and/or authorized representatives of employees, HHEs are usually initiated through NIOSH's regional representatives, although scientists from other facilities are often involved.

Requests for HHEs should be addressed to NIOSH as follows:

1. Requests from *general industry:* Hazard Evaluations and Technical Assistance Branch, Division of Surveillance, Hazards Evaluations and Field Studies, NIOSH, 4676 Columbia Parkway, Cincinnati, OH 45226.
2. Requests from *mining industry:* Environmental Investigations Branch, Division of Respiratory Disease Studies, NIOSH, 944 Chestnut Ridge Road, Morgantown, WV 26505.

Requests for HHEs should be submitted in writing and signed by either the employer or an authorized representative of employees.

Each request should contain:

- the requester's name, address, and telephone number
- the name and address of the place of employment where the substance or physical agent is normally found
- the specific process or type of work which is the source of the substance or physical agent, or in which the substance or physical agent is used
- details of the conditions or circumstances which prompted the request
- a statement if the requester is not the employer
- a statement indicating whether or not the name(s) of the requester or those persons who have authorized the requester to represent them may be revealed to the employer by NIOSH
- the following supplementary information if known to the requester:
  1. identity of each substance or physical agent involved
  2. the trade name, chemical name, and manufacturer of each substance involved
  3. whether the substance or its container or the source of the physical agent has a warning label
  4. the physical form of the substance or physical agent, number of people exposed, length of exposure (hours per day), and occupations of exposed employees

Upon receipt of a request for health hazard evaluation, NIOSH determines whether or not there is reasonable cause to justify conducting an investigation. If NIOSH determines that an investigation is not justified, the requester is notified in writing of the decision.

Advance notice of visits to the place of employment may be given. Prior to beginning an investigation, NIOSH officers present their credentials to the owner, operator, or manager of the place of employment. The officer explains the nature, purpose, and scope of the investigation and the records which they wish to review. NIOSH officers then collect environmental samples and samples of substances and/or measurements of physical agents, and/or take or obtain photographs related to the purpose of the investigation. Other reasonable investigative techniques may be employed, including medical examinations of employees (with their consent) and privately questioning any employer, owner or operator, agent, or employee. Employers have the opportunity to review photographs taken or obtained for the purpose of identifying those which contain or might reveal a trade secret.

If during the course of or as a result of an investigation the NIOSH officer believes that there is a reasonable basis for an allegation of an imminent danger, NIOSH will immediately advise the employer and those employees who appear to be in immediate danger.

Upon conclusion of an investigation, NIOSH will make a determination concerning the potentially toxic or hazardous effects of each substance or physical agent investigated as a result of the request for health hazard evaluation. Copies of the determination are mailed to the employer and authorized employee representatives, and must be posted for 30 days at or near the workplace of affected employees.

5. ACCREDITED LABORATORIES

Accredited laboratories are those that have been accredited by the American Industrial Hygiene Association (AIHA), and those which have been rated proficient by NIOSH for analyzing air samples for lead, cadmium, zinc, silica, asbestos, and organic solvents.

AIHA accreditation involves a site visit to the laboratory every three years to ascertain fulfillment of criteria for personnel qualifications, quality control, equipment, facilities, records, and methods. All AIHA-accredited laboratories must participate in the NIOSH Proficiency Analytical Testing (PAT) Program. To test proficiency, NIOSH checks accuracy of a lab's analysis of samples with known quantities of contaminants.

For a listing of accredited laboratories in your area, contact:

Lab Accreditation
American Industrial Hygiene Association
475 Wolf Ledges Parkway
Akron, OH 44311-1087
(216) 762-7291

6. CLINICAL FACILITIES FOR EVALUATING OCCUPATIONAL ILLNESS

There are a number of clinical facilities that provide occupational health services to workers. In order to locate one in your area, contact:

Association of Occupational and Environmental Clinics
P.O. Box 5214
Takoma Park, Maryland 20912-0214
(203) 785-5885

In New Jersey, contact:

New Jersey State Department of Health
Division of Occupational and Environmental Health
CN 360
Trenton, NJ 08625
(609) 984-1863

Environmental and Occupational Medicine Clinic
Department of Environmental and Community Medicine
University of Medicine and Dentistry of New Jersey
Robert Wood Johnson Medical School
675 Hoes Lane
Piscataway, NJ 08854-5635
(201) 463-4771

7. EMPLOYEE ASSISTANCE PROGRAMS

All employees, regardless of position in an organization, face a variety of problems in their daily lives. Usually, the problems can be worked out. However, sometimes the problems become too much to handle and they affect personal satisfaction, family relations, performances at work, and even health. When this occurs, professional help is often needed to resolve the problems. Without proper attention, these problems usually become worse and the consequences are often unpleasant and expensive. Employee Assistance Programs provide free and confidential professional assistance to help employees and their families resolve problems that affect their personal lives or job performance. They are voluntary programs designed to encourage individuals to seek help on their own at the earliest stage of a problem.

The programs are employer-sponsored. The employer retains the services of a qualified counseling resource that specializes in the assessment of personal problems. Counseling covers a variety of problems, including marital difficulties, financial or legal problems, emotional difficulties, or problems caused by alcohol or drug abuse.

The Occupational Section of the Division of Alcoholism of the New Jersey State Department of Health serves to promote the awareness, development, and utilization of Employee Assistance Programs in industry to resolve personal problems which affect job performance. This organization provides technical assistance to companies/unions in the fundamentals of establishing Employee Assistance Programs, and can assist in the provision of supervisory training or staff orientation activities. For further information, contact:

New Jersey State Department of Health
Occupational Services
Division of Alcoholism
CN 362
Trenton, NJ 08625-0362
(609) 292-0729

The Division of Occupational Health within the Department of Environmental and Community Medicine of UMDNJ–Robert Wood Johnson Medical School offers consultation services regarding the establishment and operation of Employee Assistance Programs for industry. Faculty within the division provide employee assistance services to other organizations and are also conversant in the academic literature in this field. Therefore, they are in a unique position to evaluate the services needed by a particular organization and to help that organization discern who may best meet those needs. For further information, contact:

Employee Assistance Program
Division of Occupational Health
Department of Environmental and Community Medicine
University of Medicine and Dentistry of New Jersey
Robert Wood Johnson Medical School
675 Hoes Lane
Piscataway, NJ 08854-5635
(201) 463-4675

8. NEW JERSEY'S WORKERS' COMPENSATION LAW

This section was adapted from "A Workers' Guide to the New Jersey Workers' Compensation Law," Labor Education Center, Institute of Management and Labor Relations, Rutgers, The State University of New Jersey.

A worker suffering from a job-related injury or occupational disease is automatically entitled to certain benefits under the New Jersey Workers' Compensation Act. The benefits include:

- medical treatment
- temporary compensation while unable to work
- total permanent disability payments for workers unable to resume any type of work
- partial permanent disability for workers able to work but with some lasting effect from the injury or disease
- death and funeral benefits for burial expenses of a deceased worker
- dependency benefits for certain dependent relatives of a deceased worker

Every type of job-related injury and disease is covered by the law. Specific procedures have been developed for filing and processing claims while waiting to collect benefits; resources are available to injured and ill workers for medical evaluation, education, and rehabilitation.

The entire workers' compensation law can be found in New Jersey Statutes Annotated (N.J.S.A.) 34:15–1 through 15–127. The law is administered by:

New Jersey Department of Labor
Division of Workers' Compensation
CN 381
Trenton, NJ 08625
(609) 292-2414

A free copy of the law can be obtained by writing to the division. There are also 13 district hearing offices throughout the state.

For more in-depth information about the law, the booklet, "A Workers' Guide to the New Jersey Workers' Compensation Law," is available for a small fee from:

Labor Education Center
IMLR
Ryders Lane
New Brunswick, NJ 08903
(201) 932-9502

9. SMALL BUSINESS ADMINISTRATION AND *SCORE* OFFICES

The Small Business Administration (SBA) provides information and assistance to small businesses. Their business loan program supplies loans which may be used for improving occupational safety and health conditions in the workplace.

The national office for the SBA can be contacted at (800) 368-5855. The SBA offices in New Jersey are:

Military Park Building
60 Park Place, 4th Floor
Newark, NJ 07102
(201) 645-2434

2600 Mt. Ephraim Avenue
Camden, NJ 08104
(609) 757-5183

The SBA also sponsors the Service Corps of Retired Executives (SCORE), which is a volunteer organization made up of experienced businessmen and women who share their knowledge with any small business manager who needs help. Their approach is confidential and person-to-person and there is no charge for this service.

The national SCORE office is located at:

SCORE
1129 20th Street, NW
Suite 410
Washington, DC 20036
(202) 653-6279

New Jersey SCORE Locations

Somerset County College
Route 28 & Lamington Road
North Branch, NJ 08876
(201) 526-1200 ext. 377/316

Small Business Administration
60 Park Place
Newark, NJ 07102
(201) 645-3982

Fairleigh Dickinson University
285 Madison Avenue
Madison, NJ 07940
(201) 593-8850

Wayne Chamber of Commerce
Broadway Bank & Trust Building

2055 Hamburg Turnpike
Wayne, NJ 07470
(201) 831-7788

County College of Morris
Center Grove Road
Randolph, NJ 07869
(201) 361-5000 ext. 584

Monmouth County Library
Route 35
Shrewsbury, NJ 07701
(201) 842-5995

Mainland Chamber of Commerce
1520 S. Main Street
Pleasantville, NJ 08232
(609) 646-0777

Ocean County College
Hooper Avenue
Toms River, NJ 08753
(201) 255-0404

Union Chamber of Commerce
2165 Morris Avenue
Union, NJ 07083
(201) 688-2777

New Brunswick Career Preparation
  Center
170 French Street
New Brunswick, NJ 08901
(201) 249-6209

Jersey City State College
90 Audubon Avenue
Jersey City, NJ 07305
(201) 547-3005

Sussex County Economic
  Development Commission
55-57 High Street
Newton, NJ 07860
(201) 383-1217

Clifton Community Center
1232 Main Avenue

Clifton, NJ 07011
(201) 470-2246

Brookdale Community College
Newman Springs Road
Lincroft, NJ 07738
(201) 842-1900 ext. 573

Western Monmouth Chamber of
  Commerce
49 E. Main Street
Freehold, NJ 07728
(201) 462-3030

Middlesex County College
Woodbridge Avenue
Edison, NJ 08818
(201) 548-6000 ext. 350

Arts & Sciences Museum
327 Ridgewood Avenue
Paramus, NJ 07625
(201) 447-7155

Small Business Administration
2600 Mt. Ephraim Avenue
Camden, NJ 08104
(201) 757-5184

10. SMALL BUSINESS RESOURCES IN NEW JERSEY

New Jersey Small Business Development Center

The New Jersey Small Business Development Center (NJSBDC), a non-profit statewide network, provides comprehensive small business management and technical assistance to small business owners and prospective owners throughout New Jersey.

NJSBDC links the resources of the U.S. Small Business Administration and the state government with those of the private sector, Rutgers Graduate School of Management, and six other SBDCs to provide a range of services to those who wish to strengthen or expand their businesses. Services available to small business owners throughout New Jersey include training and continuing education, counseling, and seminars and conferences.

New Jersey Small Business Development Center
Graduate School of Management
Rutgers, The State University of New Jersey
180 University Avenue
Newark, NJ 07102
(201) 648-5950

New Jersey Small Business Development Centers:

Atlantic Community College
Small Business Development Center
1535 Bacharach Boulevard
Atlantic City, NJ 08401
(609) 343-4810

Brookdale Community College
Small Business Development Center
Newman Springs Road
Lincroft, NJ 07738
(201) 842-1900 ext. 551

Mercer County Community College
Small Business Development Center
P.O. Box B

Trenton, NJ 08690
(609) 586-4800 ext. 469

Rutgers, The State University of
   New Jersey, Campus at Camden
Small Business Development Center
Victor Building
Point and Pearl Streets
Camden, NJ 08102
(609) 757-6221

Rutgers, The State University of
   New Jersey, Campus at Newark
Small Business Development Center
180 University Avenue
Newark, NJ 07102
(201) 648-5950

Business counseling is also available at the Raritan Valley Regional Chamber of Commerce in New Brunswick, (201) 545-3300.

New Jersey Office of Small Business Assistance

Office of Small Business Assistance
Division of Development for Small, Women, and Minority Businesses
Room 404
1 W. State Street
CN 823
Trenton, NJ 08625
(609) 984-4442

The Office of Small Business Assistance advises and encourages small, women-owned, and minority-owned businesses in matters related to establishing and operating a small business in the State of New Jersey. General managerial, financial, and economic development assistance is offered to companies with 500 or fewer employees. However, certain special projects and programs are available to New Jersey-based businesses with 100 or fewer full-time employees or to New Jersey-based women-owned and minority-owned businesses.

Some of the services include:

- approving or disapproving vendors for competition under the New Jersey Small Business Set-Aside Act
- promoting the procurement of contracts for small, women-owned, and minority-owned businesses
- informing businesses of the procedures for applying for inclusion on the state's Bidder's List and the Set-Aside Bidder's List
- providing a central resource for small businesses in their dealing with federal, state, and local governments, including providing information regarding government requirements and legislation affecting small, women-owned, and minority-owned businesses
- establishing a loan referral program and providing financial counseling for small, women-owned, and minority-owned businesses
- coordinating managerial and technical assistance in a systematic manner, using available resources within the state, including (but not restricted to) small business development centers, academic institutions with graduate business programs, and minority development offices

New Jersey Business Libraries

The Business Library of the Newark Public Library
34 Commerce Street
Newark, NJ 07102
(201) 733-7779 (Reference)

The Business Library functions as an information resource, complementing the libraries of the various educational institutions in the state. The staff is available to answer questions or to suggest other sources of information about any aspect of business. Accounting, advertising, banking, marketing, business administration, office management, insurance, investments, etc. are covered. There is a comprehensive collection of circulating books, periodicals, and information file material for use by the general public and members of the business community.

In addition to data on business and industry in the United States and abroad, the Business Library also covers New Jersey and, in particular, the

Newark area. The Business Library is open 9:00 am–5:30 pm Monday through Friday.

Rutgers University Libraries

The Rutgers libraries are available to the public as well as students.

John Cotton Dana Library
185 University Avenue
Newark, NJ 07102
(201) 648-5901

The Business Collection located in this library includes books and periodicals in business, accounting and economics, management, and administration. Corporate annual reports and 10k reports are available. A specialist is usually available 9:00 am–9:00 pm Monday through Friday, and most weekends during the academic year.

Institute of Management and Labor Relations Library
Ryders Lane and Clifton Avenue
New Brunswick, NJ 08903
(201) 932-9513

The Industrial Relations and Labor Collection is an extensive, specialized collection in all phases of industrial relations. Reports, services, journals, government documents, and books are available. The library interacts extensively statewide with labor and management organizations, negotiators and mediators. Specialists are available 9:00 am–9:00 pm Monday through Thursday, 9:00 am–5:00 pm Friday, year-round.

APPENDIX A

Resources for Chapter One

OCCUPATIONAL HEALTH RESOURCES

Selected Publications

Medical Publications, Inc. publishes the journal *Occupational Safety and Health* 13 times a year ($36 for yearly subscription). It also publishes the *1987/88 Occupational Health and Safety Purchasing Sourcebook,* a comprehensive guide for employers. This directory includes products and services, company profiles, trade/brand names and toll-free numbers ($29.95).

Occupational Health and Safety
Medical Publications, Inc.
225 N. New Road
Waco, TX 76710
(817) 776-9000

The Bureau of National Affairs, Inc. publishes the *Occupational Safety and Health Reporter,* a weekly update of legislation. (Call for price.)

The Bureau of National Affairs, Inc.
9435 Key West Avenue
Rockville, MD 20850
(800) 372-1033

National Nonprofit Organizations

* American Conference of Governmental Industrial Hygienists (ACGIH)
  6500 Glenway Avenue, Building D-7

Cincinnati, OH 45211
(513) 661-7881

ACGIH publishes a nationally recognized source of information about workplace hazardous chemicals: "Threshold Limit Values and Biological Exposure Indices" ($4.50 prepaid).

* American Industrial Hygiene Association (AIHA)
  475 Wolf Ledges Parkway
  Akron, OH 44311-1087
  (216) 762-7294

AIHA publishes the Hygienic Guides series with health and physical hazard information on 200 common industrial chemicals (individual copies $5 prepaid, $1.50 shipping and handling). Also, AIHA supplies listings of accredited laboratories. (If requesting by mail, add "Attention: Lab Accreditation" to address.)

* American National Standards Institute (ANSI)
  1430 Broadway
  New York, NY 10018
  (212) 354-3300

ANSI publishes health and safety standards for numerous industrial processes and toxic substances. ANSI concentrates on technical recommendations for improving the work environment.

* National Institute for Occupational Safety and Health (NIOSH)
  4676 Columbia Parkway
  Cincinnati, OH 45226

  (The national office also serves as the regional office for the central United States.)

  Telephone numbers:

  (800) 356-4674 (technical information)
  (513) 533-8328 (more detailed technical information)
  (513) 533-8287 (publications)
  (513) 533-8221 (Division of Training and Manpower Development)
  (513) 841-4428 (Division of Surveillance, Hazard Evaluations, and Field Studies)

NIOSH is the principal federal agency engaged in research and is responsible for identifying occupational safety and health hazards and for recommending changes in the regulations limiting them. It also has obligations for training occupational health manpower.

The NIOSH Regional Offices are:

Boston Region (covers northeastern United States)
1401 JFK Federal Building
USDHS/CDC/NIOSH
Public Health Service
Boston, MA 02203
(617) 565-1439

Atlanta Region (covers southeastern United States)
Department of Health and Human Services
Public Health Service
Division of Preventive Health Services
Suite 1110
101 Mariatta Tower
Atlanta, GA 30323
(404) 331-2396

Denver Region (covers western United States)
U.S. Public Health Service
NIOSH
1961 Stout Street
Denver, CO 80294
(303) 844-6166

• National Safety Council (NSC)
  444 North Michigan Avenue
  Chicago, IL 60660
  (312) 527-4800

NSC publishes the *Small Business Safety and Health Manual,* which is designed to help small business owners/managers achieve compliance with OSHA guidelines (single copy, for nonmembers $13.50). Also available is *Protecting Worker's Lives: A Safety and Health Guide for Unions,* which contains information on health hazards, fire prevention, workplace inspection and investigation, organizing committees, etc. (single copy, for nonmembers $19.50). A catalog is available.

States may also have a local safety council. In New Jersey, the council is located at the following address:

New Jersey Safety Council
6 Commerce Drive
Cranford, NJ 07016
(201) 272-7712

Committees on Occupational Safety and Health (COSH)

COSH groups are independent local organizations of workers, local unions, and health, safety, and legal professionals concerned with industrial health hazards and worker protection. They provide information to employees and unions on a regional basis.

COSH Groups

Alaska

Alaska Health Project
417 W. 7th Avenue, Suite 101
Anchorage, AL 99501
907-276-2864

California

BACOSH (San Francisco Bay Area)
LOPH, Institute of Industrial
  Relations
University of California
2521 Channing Way
Berkeley, CA 94720
415-482-1095

LACOSH (Los Angeles COSH)
2501 S. Hill Street
Los Angeles, CA 90007
213-749-6161

Sacramento COSH
c/o Fire Fighters Local 522
3101 Stockton Boulevard
Sacramento, CA 95820
916-444-8134

SCCOSH (Santa Clara Center for
  OSH)
760 North 1st Street
San Jose, CA 95110
408-998-4050

Connecticut

ConnectiCOSH
130 Huyhope Street
Hartford, CT 06106
203-549-1877

District of Columbia

Alice Hamilton Occupational Health
  Center
410 Seventh Street, SE
Washington, DC 20003
202-543-0005

Illinois

CACOSH (Chicago COSH)
33 East Congress Expressway
  Suite 723
Chicago, IL 60605
312-939-2104

Maine

Maine Labor Group on Health, Inc.
Box V
Augusta, ME 04330
207-289-2770

Maryland

MaryCOSH
325 East 25th Street
Baltimore, MD 21218
301-467-3666

Massachusetts

MassCOSH (Massachusetts COSH)
718 Huntington Avenue
Boston, MA 02115
617-277-0097

Michigan

SEMCOSH (Southeast Michigan
 COSH)
1550 Howard Street
Detroit, MI 48216
313-961-3345

New York

ALCOSH (Allegheny Council on
 Occupational Safety and Health)
210 W. 5th Street
Jamestown, NY 14701
716-488-0720

CNYCOSH (Central New York
 COSH)
615 W. Genessee Street
Syracuse, NY 13204
315-437-9401

NYCOSH (New York COSH)
275 Seventh Avenue, 25th Fl.
New York, NY 10001
212-627-3900

ROCOSH (Rochester COSH)
502 Lyell Avenue, Suite #1

Rochester, NY 14606
716-458-8553

WNYCOSH (Western New York
 COSH)
450 Grider Street
Buffalo, NY 14215
716-897-2110

North Carolina

NCOSH (North Carolina OSH
 Project)
P.O. Box 2514
Durham, NC 27705
919-286-9249

Ohio

ORVCOSH (Ohio River Valley
 COSH)
35 E. 7th Street, Suite 200
Cincinnati, OH 45202
513-421-1849

Pennsylvania

PHILAPOSH (Philadelphia Project
 OSH)
511 N. Broad Street, Suite 900
Philadelphia, PA 19123
215-386-7000

Rhode Island

RICOSH (Rhode Island COSH)
340 Lockwood Street
Providence, RI 02907
401-751-2015

Tennessee

TNCOSH (Tennessee COSH)
1515 E. Magnolia, Suite 406

Knoxville, TN 37917
615-525-3147 or 615-525-5090

Wisconsin

WISCOSH (Wisconsin COSH)
1334 S. 11th Street
Milwaukee, WI 53204
414-643-0928

Canada

WOSH (Windsor OSH Project)
1109 Tecumseh Road East
Windsor, Ontario N8W2T1, Canada
519-254-4192

VANCOSH
616 East 10th Avenue
Vancouver, BC V5T2A5, Canada

New Jersey Agencies

New Jersey State Department of Health (DOH)
Occupational Health Service
CN 360
Trenton, NJ 08625-0360
(609) 984-1863

DOH provides occupational safety and health information and materials relating to private sector workplaces. Staff industrial hygienists perform workplace inspections in response to employer and employee requests and conduct NIOSH-assigned Health Hazard Evaluations in the State of New Jersey.

In addition, the Public Employee Occupational Safety and Health Project provides occupational health information, education, and training materials to public employees and employers and inspects public workplaces for violations of Public Employees Occupational Safety and Health Act (PEOSHA) health standards.

The Occupational Disease Clinic provides medical evaluations for individuals or groups with illness or injury due to occupational or environmental exposures. Call above number to arrange appointment.

Available from DOH is the *Occupational Health Resource Guide,* a 250-page guide for locating occupational health organizations, written and audiovisual resources, and data bases. Over 1000 entries are included, covering many specific occupational health topics.

New Jersey Department of Labor
CN 054
Trenton, NJ 08625-0054
(609) 984-3507 (OSHA Consultation Services)
(609) 292-7036 (Office of Public Employees Safety)
(800) 624-1644 (PEOSHA Complaint Hotline)

Through OSHA Consultation Services, consultants perform plant inspections and help industries to design methods of hazard control and means of meeting OSHA requirements. (For more information, see Chapter 5, Section 2.)

New Jersey Local/County Occupational Health Programs

City of Elizabeth
50 Winnfield Scott Plaza
Elizabeth, NJ 07201
(201) 820-4049 or -4060
(covers Elizabeth and Union County)

Occupational Health Care
Consortium of Northern NJ
Paterson Health Dept.
176 Broadway
Paterson, NJ 07505
(201) 881-3914
(covers Passaic County)

Hudson Regional Health
   Commission
215 Harrison Avenue
Harrison, NJ 07029
(201) 485-7001
(covers Hudson County)

Atlantic County Department of
   Health
201 South Shore Road
Northfield, NJ 08225

(609) 645-7700 ext. 4372
(covers Atlantic County except
   Atlantic City)

Bergen County Department of
   Health Services
327 Ridgewood Avenue
Paramus, NJ 07652-4895
(201) 599-6100
(covers Bergen County)

Clifton Health Department
900 Clifton Avenue
Clifton, NJ 07011
(201) 470-5758
(covers City of Clifton)

Atlantic City Division of Health
35 South Illinois Avenue
Atlantic City, NJ 08401
(609) 347-5671
(covers Atlantic City; no formal
   program, responds to calls and
   assigns sanitary inspectors)

At this printing, not all areas have occupational health programs; however, the State of New Jersey requires all Health Departments to have occupational health programs by January 1989.

APPENDIX B

Resources for Chapter Two

1. OCCUPATIONAL SAFETY AND HEALTH RESOURCES

(For additional resources, see Appendix D, Section 1.)

How to Obtain OSHA Standards

The *Federal Register* is one of the best sources of information on standards, since all OSHA standards are published in the *Federal Register* when adopted, as are all amendments, corrections, insertions, or deletions. The *Federal Register* is available in many public libraries. Annual subscriptions are available from the Superintendent of Documents, U.S. Government Printing Office, Washington, DC 20402 ($340 per year).

Each year the Office of the *Federal Register* publishes all current regulations and standards in the *Code of Federal Regulations* (CFR), available at many libraries and from the Government Printing Office. OSHA's regulations are collected in Title 29 of the CFR, Parts 1900–1999. Parts 1900–1910 (General Industry) and 1926 (Construction) are available from local OSHA offices (individual copies free of charge).

The OSHA Subscription Service was developed to assist the public in keeping current with OSHA standards. The service is available from the Superintendent of Documents (see above address) only, and is not available from OSHA or from the Department of Labor. Individual volumes of the OSHA Subscription Service are available as follows:

Volume I. General Industry Standards and Interpretations (includes agriculture) ($97 per year)

Volume II. Maritime Standards and Interpretations ($29 per year)

Volume III. Construction Standards and Interpretations ($29 per year)

Volume IV. Other Regulations and Procedures ($71 per year)

Volume V. Field Operations Manual ($28 per year)

Volume VI. Industrial Hygiene Field Operations Manual ($38 per year)

Selected Publications

General

Lowrys' Handbook of Right-to-Know and Emergency Planning explains in nonlegalistic terms what is required and how to comply with the Hazard Communication Standard and SARA Title III. It can be purchased from:

Lewis Publishers, Inc.
121 South Main Street
P.O. Drawer 519
Chelsea, Michigan 48118
(800) 525-7894

Selected OSHA Publications

"All About OSHA," OSHA 2056 (1985 revised). A good brief summary of the OSHA law, coverage, standards, and penalties.

"OSHA Handbook for Small Businesses," OSHA 2209 (1979 revised). Handbook to assist small business employers to meet their legal requirements under the OSHAct.

OSHA Offices

National Office

U.S. Department of Labor
Occupational Safety and Health Administration
3rd & Constitution Avenue, N.W.
Washington, DC 20210

Regional Offices

(For a listing of the states by standard federal regions, see Appendix E, Section 1.)

Region I (CT,[2] ME, MA, NH, RI, VT[1])
16–18 North Street
1 Dock Square Building, 4th Floor
Boston, MA 02109
(617) 565-1145

Region II (NY,[2] NJ, PR[1])
201 Varick Street
New York, NY 10014
(212) 337-2378

Region III (DE, DC, MD,[1] PA, VA,[1] WV)
Gateway Building, Suite 2100
3535 Market Street
Philadelphia, PA 19104
(215) 596-1201

Region IV (AL, FL, GA, KY,[1] MS, NC,[1] SC,[1] TN[1])
1375 Peachtree Street, N.E., Suite 587
Atlanta, GA 30367
(404) 347-3573

Region V (IL, IN,[1] MN,[1] MI,[1] OH, WI)
230 South Dearborn Street
32nd Floor, Room 3244
Chicago, IL 60604
(312) 353-2220

Region VI (AR, LA, NM,[1] OK, TX)
525 Griffin Square, Room 602
Dallas, TX 75202
(214) 767-4731

Region VII (IA,[1] KS, MO, NE)
911 Walnut Street, Room 406
Kansas City, MO 64106
(816) 374-5861

Region VIII (CO, MT, ND, SD, UT,[1] WY[1])
Federal Building, Room 1576
1961 Stout Street
Denver, CO 80294
(303) 844-3061

Region IX (AZ,[1] CA,[1] HI,[1] NV)
71 Stevenson Street, 4th Floor
San Francisco, CA 94105
(415) 995-5672

Region X (AK,[1] ID, OR,[1] WA[1])
Federal Office Building, Room 6003
909 First Avenue
Seattle, WA 98174
(206) 442-5930

[1]State operates an approved enforcement program in both the public and private sectors.
[2]State operates a public employee-only program (Connecticut and New York).

Area Offices

Alabama

Occupational Safety and Health
  Admin.
2047 Canyon Road—Todd Mall
Birmingham, AL 35216
(205) 731-1534

Alaska

Occupational Safety and Health
  Admin.
Federal Building
701 "C" Street, Box 29
Anchorage, AK 99513
(907) 271-5152

Arizona

Occupational Safety and Health
  Admin.
3221 North 16th Street, Suite 100
Phoenix, AZ 85016
(602) 241-2006

Arkansas

Occupational Safety and Health
  Admin.
Savers Building, Suite 828
320 West Capitol Avenue
Little Rock, AR 72201
(501) 378-6291

California

Occupational Safety and Health
  Admin.
400 Oceangate, Suite 530
Long Beach, CA 90802
(213) 514-6387

Occupational Safety and Health
  Admin.
2422 Arden Way, Suite A-1
Sacramento, CA 95825
(916) 646-9220

Occupational Safety and Health
  Admin.
7807 Convoy Court, Suite 160
San Diego, CA 92111
(619) 569-9071

Occupational Safety and Health
  Admin.
950 S. Bascom, Suite 3120
San Jose, CA 95128
(408) 291-4600

Occupational Safety and Health
  Admin.
801 Ygnacio Valley Rd., Room 205
Walnut Creek, CA 94596-3823
(415) 943-1973

Occupational Safety and Health
  Admin.
100 N. Citrus Avenue, Suite 240
West Covina, CA 91791
(818) 915-1558

Colorado

Occupational Safety and Health
  Admin.
Tremont Center, 1st Floor
333 West Colfax
Denver, CO 80204
(303) 844-5285

Connecticut

Occupational Safety and Health
   Admin.
Federal Office Building
450 Main Street, Room 508
Hartford, CT 06103
(203) 240-3152

Florida

Occupational Safety and Health
   Admin.
299 East Broward Boulevard
Fort Lauderdale, FL 33301
(305) 527-7292

Occupational Safety and Health
   Admin.
3100 University Boulevard South
Jacksonville, FL 32207
(904) 791-2895

Occupational Safety and Health
   Admin.
700 Twiggs Street, Room 624
Tampa, FL 33602
(813) 228-2821

Georgia

Occupational Safety and Health
   Admin.
Building 10, Suite 33
La Vista Perimeter Officer Park
Tucker, GA 30084
(404) 331-4767

Hawaii

Occupational Safety and Health
   Admin.
300 Ala Moana Boulevard, Suite
   5122

P.O. Box 50072
Honolulu, HI 96850
(808) 541-2685

Idaho

Occupational Safety and Health
   Admin.
Federal Bldg./USCH, Room 324
550 West Fort Street, Box 007
Boise, ID 83724
(208) 334-1867

Illinois

Occupational Safety and Health
   Admin.
344 Smoke Tree Business Park
North Aurora, IL 60542
(312) 896-8700

Occupational Safety and Health
   Admin.
1600 167th Street, Suite 12
Calumet City, IL 60409
(312) 891-3800

Occupational Safety and Health
   Admin.
6000 W. Touhy Avenue
Niles, IL 60648
(312) 631-8200

Occupational Safety and Health
   Admin.
2001 West Willow Knolls Rd., Suite
   101
Peoria, IL 61614
(309) 671-7033

Indiana

Occupational Safety and Health
   Admin.

U.S. Post Office and Courthouse
46 East Ohio Street, Room 423
Indianapolis, IN 46204
(317) 269-7290

Iowa

Occupational Safety and Health
  Admin.
210 Walnut Street, Room 815
Des Moines, IA 50309
(515) 284-4794

Kansas

Occupational Safety and Health
  Admin.
216 N. Waco, Suite B
Wichita, KS 67202
(316) 269-6644

Kentucky

Occupational Safety and Health
  Admin.
John C. Watts Federal Building,
  Room 108
330 W. Broadway
Frankfort, KY 40601
(502) 227-7024

Louisiana

Occupational Safety and Health
  Admin.
2156 Wooddale Boulevard
Hoover Annex, Suite 200
Baton Rouge, LA 70806
(504) 389-0474

Maine

Occupational Safety and Health
  Admin.

U.S. Federal Building
40 Western Avenue, Room 121
Augusta, ME 04330
(207) 622-8417

Maryland

Occupational Safety and Health
  Admin.
Federal Building, Room 1110
Charles Center, 31 Hopkins Plaza
Baltimore, MD 21201
(301) 962-2840

Massachusetts

Occupational Safety and Health
  Admin.
1550 Main Street, Room 532
Springfield, MA 01103-1493
(413) 785-0123

Occupational Safety and Health
  Admin.
400–2 Totten Pond Road, 2nd Floor
Waltham, MA 02154
(617) 647-8681

Michigan

Occupational Safety and Health
  Admin.
300 E. Michigan, Room 202
Lansing, MI 48933
(517) 377-1892

Minnesota

Occupational Safety and Health
  Admin.
110 South 4th Street, Room 425
Minneapolis, MN 55401
(612) 348-1994

Mississippi

Occupational Safety and Health
 Admin.
Federal Building, Suite 1445
100 West Capitol Street
Jackson, MS 39269
(601) 965-4606

Missouri

Occupational Safety and Health
 Admin.
911 Walnut Street, Room 2202
Kansas City, MO 64106
(816) 374-2756

Occupational Safety and Health
 Admin.
4300 Goodfellow Boulevard,
 Building 105E
St. Louis, MO 63120
(314) 263-2749

Montana

Occupational Safety and Health
 Admin.
19 N. 25th Street
Billings, MT 59101
(406) 657-6649

Nebraska

Occupational Safety and Health
 Admin.
Overland-Wolf Building, Room 100
6910 Pacific Street
Omaha, NE 68106
(402) 221-3182

New Hampshire

Occupational Safety and Health
 Admin.
Federal Building, Room 334
55 Pleasant Street
Concord, NH 03301
(603) 225-1639

New Jersey

Occupational Safety and Health
 Admin.
Plaza 35, Suite 205
1030 Saint Georges Avenue
Avenel, NJ 07001
(201) 750-3270

Occupational Safety and Health
 Admin.
2101 Ferry Avenue, Room 403
Camden, NJ 08104
(609) 757-5181

Occupational Safety and Health
 Admin.
2 E. Blackwell Street
Dover, NJ 07801
(201) 361-4050

Occupational Safety and Health
 Admin.
500 Route 17 South, 2nd Floor
Hasbrouck Heights, NJ 07604
(201) 288-1700

New Mexico

Occupational Safety and Health
 Admin.
320 Central Avenue, S.W., Suite 13
Albuquerque, NM 87102
(505) 766-3411

New York

Occupational Safety and Health
  Admin.
Leo W. O'Brien Federal Building
Clinton Avenue and N. Pearl Street
Room 132
Albany, NY 12207
(518) 472-6085

Occupational Safety and Health
  Admin.
5360 Genesee Street
Bowmansville, NY 14026
(716) 684-3891

Occupational Safety and Health
  Admin.
136-21 Roosevelt Avenue
Flushing, NY 11354
(718) 445-5005

Occupational Safety and Health
  Admin.
90 Church Street, Room 1405
New York, NY 10007
(212) 264-9840

Occupational Safety and Health
  Admin.
100 S. Clinton Street, Room 1267
Syracuse, NY 13260
(315) 423-5188

Occupational Safety and Health
  Admin.
990 Westbury Rd.
Westbury, NY 11590
(516) 334-3344

North Carolina

Occupational Safety and Health
  Admin.

Century Station, Room 104
300 Fayetteville Street Mall
Raleigh, NC 27601
(919) 856-4770

North Dakota

Occupational Safety and Health
  Admin.
Federal Building, Room 348
P.O. Box 2439
Bismarck, ND 58501
(701) 255-4011 ext. 521

Ohio

Occupational Safety and Health
  Admin.
Federal Office Building, Room 4028
550 Main Street
Cincinnati, OH 45202
(513) 684-3784

Occupational Safety and Health
  Admin.
Federal Office Building, Room 899
1240 East Ninth Street
Cleveland, OH 44199
(216) 522-3818

Occupational Safety and Health
  Admin.
Federal Office Building, Room 634
200 N. High Street
Columbus, OH 43215
(614) 469-5582

Occupational Safety and Health
  Admin.
Federal Office Building, Room 734
234 N. Summit Street
Toledo, OH 43604
(419) 259-7542

Oklahoma

Occupational Safety and Health
Admin.
420 West Main Place, Suite 725
Oklahoma City, OK 73102
(405) 231-5351

Oregon

Occupational Safety and Health
Admin.
1220 S.W. Third Street, Room 640
Portland, OR 97204
(503) 221-2251

Pennsylvania

Occupational Safety and Health
Admin.
Progress Plaza
49 N. Progress Avenue
Harrisburg, PA 17109
(717) 782-3902

Occupational Safety and Health
Admin.
U.S. Custom House, Room 242
Second and Chestnut Street
Philadelphia, PA 19106
(215) 597-4955

Occupational Safety and Health
Admin.
1000 Liberty Avenue, Room 2236
Pittsburgh, PA 15222
(412) 644-2903

Occupational Safety and Health
Admin.
Penn Place, Room 2005
20 North Pennsylvania Avenue
Wilkes-Barre, PA 18701
(717) 826-6538

Puerto Rico

Occupational Safety and Health
Admin.
U.S. Courthouse & FOB
Carlos Chardon Avenue, Room 559
Hato Rey, PR 00918
(809) 753-4457/4072

Rhode Island

Occupational Safety and Health
Admin.
380 Westminster Mall
Room 243
Providence, RI 02903
(401) 528-4669

South Carolina

Occupational Safety and Health
Admin.
1835 Assembly Street, Room 1468
Columbia, SC 29201
(803) 765-5904

Tennessee

Occupational Safety and Health
Admin.
2002 Richard Jones Rd., Suite C-
205
Nashville, TN 37215
(615) 736-5313

Texas

Occupational Safety and Health
Admin.
611 East 6th Street, Room 303
Austin, TX 78701
(512) 482-5783

Occupational Safety and Health
Admin.
Government Plaza, Room 300
400 Mann Street
Corpus Christi, TX 78401
(512) 888-3257

Occupational Safety and Health
Admin.
2320 La Branch Street, Room 1103
Houston, TX 77004
(713) 750-1727

Occupational Safety and Health
Admin.
1425 W. Pioneer Drive
Irving, TX 75061
(214) 767-5347

Occupational Safety and Health
Admin.
Federal Building, Room 422
1205 Texas Avenue
Lubbock, TX 79401
(806) 743-7681

Utah

Occupational Safety and Health
Admin.
1781 South 300 West
Salt Lake City, UT 84115
(801) 524-5080

NIOSH

Washington

Occupational Safety and Health
Admin.
121 107th Street, N.E.
Bellevue, WA 98004
(206) 442-7520

West Virginia

Occupational Safety and Health
Admin.
550 Eagan Street, Room 206
Charleston, WV 25301
(304) 347-5937

Wisconsin

Occupational Safety and Health
Admin.
2618 North Ballard Road
Appleton, WI 54915
(414) 734-4521

Occupational Safety and Health
Admin.
2934 Fish Hatchery Rd., Suite 225
Madison, WI 53713
(608) 264-5388

Occupational Safety and Health
Admin.
Henry S. Reuss Building, Suite 1180
310 West Wisconsin Avenue
Milwaukee, WI 53203
(414) 291-3315

Selected Publication

"NIOSH Recommendations for Occupational Safety and Health Standards,"
Publications Dissemination, Division of Standards Development and Technology Transfer, NIOSH, 4676 Columbia Parkway, Cincinnati, Ohio 45226,
(513) 841-4287 (enclose self-addressed mailing label with request).

National Office

The national NIOSH office also serves as the regional office for the central United States.

NIOSH
4676 Columbia Parkway
Cincinnati, OH 45226
(800) 356-4674 (technical information)
(513) 533-8328 (more detailed technical information)
(513) 533-8287 (publications)
(513) 533-8221 (Division of Training and Manpower Development)
(513) 841-4428 (Division of Surveillance, Hazard Evaluations, and Field Studies)

Regional Offices

Boston Region (covers northeastern United States)
USDHS/CDC/NIOSH
Public Health Service
1401 JFK Federal Building
Boston, MA 02203
(617) 565-1439

Atlanta Region (covers southeastern United States)
USDHS/CDC/NIOSH
Public Health Service
Division of Preventive Health Services
Suite 1110
101 Mariatta Tower
Atlanta, GA 30323
(404) 331-2396

Denver Region (covers western United States)
USDHS/CDC/NIOSH
Public Health Service
1961 Stout Street
Denver, CO 80294
(303) 844-6166

OSHA Hazard Communication Standard: Summary

1. Are you covered by the standard?

    Yes, if you are an employer covered by OSHA (10 employees or more) or an importer or distributor of hazardous chemicals.

2. If you are a chemical manufacturer or importer:

You must evaluate (review the available information about) your chemicals to determine if they are physical or health hazards.

3. If you are a chemical manufacturer, importer, or distributor:

You must *label* hazardous chemicals with an identifier, hazard warnings, and your name and address.

You must forward *Material Safety Data Sheets* (MSDSs) with your first shipment of chemicals to a buyer.

4. All employers covered by this Standard must:
   a. Develop a *written hazard communication program* and make it available to employees, employee representatives, OSHA, and NIOSH.

   It must include:
   • labeling
   • MSDSs
   • employee training
   • list of hazardous substances in each work area
   • methods to inform employees of nonroutine hazards and of unlabeled pipes
   • methods to inform contractors about hazardous chemicals
   b. Maintain *labels* on containers and keep an MSDS for each hazardous substance in each department where it is used.
   c. Make *MSDSs* available to employees, employee representatives, OSHA, and NIOSH.
   d. *Train* employees on hazardous chemicals in their work area, after they are given a new assignment, or if new hazards are introduced.

2. NEW JERSEY RIGHT TO KNOW RESOURCES

New Jersey Right to Know Law: Summary

1. Are you covered by the law?
   • Find out your Standard Industrial Classification Code (SIC Code) by calling the New Jersey Department of Labor at (609) 292-2633, to see if you are included in the list of employers covered by the law (Chapter 2, Table 2.1).

   • If you are a *manufacturing* employer (SIC Code 20–39) or a covered *non-manufacturing* employer (see Table 2.1), you have to fill out a combined Right to Know/Title III survey, which will be supplied by DEP and will be applicable to the community provisions of the law. In this case, you do not have to comply with New Jersey's labeling or education and training requirements, but you do have to comply with OSHA's Hazard Communication Program requirements. If you are a *public agency*, you must complete a New Jersey Right to Know Survey and comply with the labeling and education and training requirements of the New Jersey RTK law.

- If you are a *research & development laboratory* in a public agency that has applied for and been granted an R&D exemption, you have to provide employee education and training, set up a communication program with the local fire department, keep Hazardous Substance Fact Sheets, and label containers (using a code or number system if desired), but you do not have to file a New Jersey Right to Know Survey.

2. *Public agencies* are required to fill out a Right to Know survey (developed by DEP and DOH) and mail it to the following agencies:

- DEP (with copy for DOH)
- county lead agency
- local police and fire departments

Private employers must fill out a combined Right to Know/Title III survey (developed by DEP and EPA) and mail it to the following agencies:

- DEP
- county lead agency
- local police and fire departments

3. The following forms must be kept in the facility in a central file, and must be made available to employees within five working days of a request:

- Right to Know survey or RTK/Title III survey
- Hazardous Substance Fact Sheets (public agencies only)
- Right to Know Hazardous Substance List

4. Inform employees of the Right to Know Law (public agencies only). Right to Know posters (provided by DOH) must be displayed.

5. Provide education and training (public agencies only):

- to current employees

- annually

- to new or reassigned employees within one month of employment or reassignment

6. *Label* (public agencies only) pipelines and *all* containers before employees open containers or within five working days of arrival of material, whichever is sooner.

7. To find unknown chemical components of mixtures:

- Request an MSDS from the manufacturer or supplier.

- If unable to identify components, submit form to DEP with documentation of "good faith" effort.

Obtaining Copies of the New Jersey Right to Know Act

Copies of the act can be obtained at no charge from the New Jersey Legislative Bill Room, State House Annex, Room 14, Trenton, NJ 08625, (609) 292-6395, by requesting Chapter 315, laws of 1983.

New Jersey State Department of Health regulations on the law (N.J.A.C. 8:59-1 through 10) can be obtained from the New Jersey State Department of Health (DOH), Right to Know Project, Occupational Disease Prevention and Information Program, CN 368, Trenton, NJ 08625, (609) 984-2202. Copies of the law are also available at no charge at the DOH.

For any questions concerning the Right to Know Act, contact the Right to Know Interagency Infoline at (609) 984-5627, the Department of Health Infoline at (609) 984-2202, or the Department of Environmental Protection at (609) 292-6714.

How to Obtain Hazardous Substance Fact Sheets

For a list of Hazardous Substance Fact Sheets available, as well as prices and postage and handling charges, contact:

New Jersey State Department of Health
Right to Know Project
CN 368
Trenton, NJ 08625
(609) 984-2202

Right to Know Governmental Information Resources

For information regarding New Jersey Right to Know law implementation and enforcement, contact the following agencies:

Right to Know Interagency Infoline: (609) 984-5627
(New Jersey Departments of Health, Environmental Protection, and Labor)

The Interagency Infoline registers complaints of violations of the law and will direct more specific questions to the proper agency.

The New Jersey State Department of Health
Right to Know Project
Occupational Disease Prevention and Information Program
CN 368
Trenton, NJ 08625-0368
Right to Know Infoline: (609) 984-2202

This agency provides information on labeling containers and setting up education and training for employees; fulfills requests for Hazardous Substance Fact Sheets and informational materials; provides technical information on health and safety concerns of specific chemical substances; and answers general inquiries relating to the Department's role in implementing and enforcing the Right to Know law in the workplace. They also provide in Spanish, upon request, survey forms, Hazardous Substance Fact Sheets, brochures, and posters. Bilingual personnel are available to answer questions in Spanish. DOH also provides "policy interpretations" for specific topics under the Right to Know law, including photocopying machines, research and development labs, consumer products, and training volunteer firefighters.

The New Jersey Department of Environmental Protection
Bureau of Hazardous Substances Information
Division of Environmental Quality
CN 405
Trenton, NJ 08625-0405
Right to Know Infoline: (609) 292-6714

This agency provides information to assist employers in the completion of the New Jersey Right to Know Survey; processes requests for surveys, survey packets, and informational materials; answers general inquiries relating to DEP's role in implementing the New Jersey Right to Know law; provides technical information and educational materials about hazardous substances in the environment; and implements the Superfund Amendments and Reauthorization Act, Title III (Community Right-to-Know).

The New Jersey Department of Labor
CN 394
Trenton, NJ 08625
Assessment of Fees: (609) 292-2989
Reclassification or Justification of SIC Codes: (609) 292-2633

This agency provides information concerning assessment of fees and requests for SIC Code reclassification and assigns SIC Codes to newly formed businesses.

The New Jersey Department of the Public Advocate
CN 850
Trenton, NJ 08625
Division of Public Interest Advocacy: (609) 292-1692
Division of Citizen Complaints: (609) 984-3131

The Division of Public Interest Advocacy represents the public interest in litigation involving the Right to Know law. The Division of Citizen Complaints handles complaints about state agency implementation of laws.

County Lead Agencies

This list includes county health departments, county clerks, and other designated health agencies responsible for maintaining a file of Right to Know Surveys and Hazardous Substance Fact Sheets for all covered employers within their counties. These files are available for public inspection.

Atlantic County Health Dept.
(609) 645-7700 ext. 4372

Bergen County Dept. of Health
Services
(201) 599-6100

Burlington County Health Dept.
(609) 265-5539

Camden County Dept. of Health
(609) 757-8600

Cape May County Dept. of Health
(609) 465-1208

Cumberland County Health Dept.
(609) 451-8000 ext. 374

Essex County Dept. of Health/
Rehabilitation
(201) 228-8319

Gloucester County Health Dept.
(609) 853-3428

Hudson Regional Health
    Commission (Hudson County)
(201) 485-7001

Hunterdon County Health
    Commission
(201) 788-1351 ext. 350

Mercer County Clerk
(609) 989-6464

Middlesex County Health Dept.
(201) 828-8100 ext. 2226

Monmouth County Health Dept.
(201) 431-7456

Morris County Clerk
(201) 829-8300

Ocean County Health Dept.
(201) 341-9700 ext. 431

Paterson Health Dept. (Passaic
    County)
(201) 881-3914

Salem County Health Dept.
(609) 769-2126

Somerset County Clerk
(201) 231-7000 ext. 7511

Sussex County Health Dept.
(201) 948-4545

Union County Bureau of
    Environmental Affairs
(201) 527-4215

Warren County Health Dept.
(201) 689-6693

Understanding and Using MSDSs

Adapted from "Art Hazard News," August 1985, Center for Occupational Hazards, 5 Beekman Street, New York, NY 10038.

MSDSs are fact sheets on the hazards of chemicals or mixtures of chemicals and are usually obtained from the chemical manufacturer, importer, or distributor. To be effective, the MSDSs should be complete, accurate, and up to date.

All MSDSs are arranged in the same format so as to avoid confusion and save time. The following is an outline of information contained in an MSDS:

Section 1 lists the product name, synonyms, and the manufacturer.

Section 2 is a breakdown of the product's hazardous components and their rough proportions. Included are Threshold Limit Values (TLVs), which are levels of airborne contaminants that many people can be exposed to without adverse effects, and Permissible Exposure Limits (PELs), which are the legal maximum concentrations in the air for chemicals, averaged over eight hours.

Section 3 includes physical information such as vapor pressure, water solubility (dissolvability), appearance, and odor.

Section 4 lists flash point, flammability, fire and explosion hazards, and proper fire extinguisher use.

Section 5 describes the known short-term health effects, symptoms of overexposure, and first aid procedures. Chronic (long-term) health effects were rarely included in the past, but OSHA's Hazard Communication Standard also requires that information.

Section 6 lists materials that should not be stored near the substance because of potential for violent reactions.

Section 7 has important information on cleaning up a spill or leak of the material and on disposing of it safely.

Section 8 covers personal protective equipment such as safety glasses, gloves, or respirators that must be worn.

Section 9 describes special handling precautions not covered in other sections.

Sometimes sections are left blank on an MSDS. This does *not* mean that there are no hazards in that area; it may just mean that the information was left out. The OSHA Hazard Communication Standard does not allow blank sections on an MSDS. When ordering a chemical, insist that a copy of the most recent MSDS come with the order. If it is not provided, consider finding another purchasing source for the chemicals.

The most complete and easiest to understand MSDSs currently available are called Hazardous Substance Fact Sheets (HSFS) and over 2000 are being produced by DOH as required by the New Jersey Right to Know Law. (See "How to Obtain Hazardous Substance Fact Sheets," above.) However, these fact sheets are for individual chemicals and not mixtures.

Dr. Daniel Marsick of OSHA, speaking to the American Chemical Society Division of Chemical Health and Safety on September 10, 1985, described several MSDS resource collections that are available in the marketplace, each with its own strengths and weaknesses. These are listed below and summarized in Table B.1.

Occupational Health Services, Inc. (OHS), 450 7th Avenue, Suite 2407, New York, NY 10123, (212) 967-1100.

Information Handling Services (IHS), 15 Inverness Way East, P.O. Box 1154, Englewood, CO 80150, (800) 525-7052.

Genium Publishing Corporation, 1145 Catalyn Street, Schenectady, NY 12303, (518) 377-8854

Table B.1.   Material Safety Data Sheet Collection Features

Selection Factors	OHS	IHS	Genium	VCH	HMIS
Number of chemicals	3000	8500	1288	867	—
Number of MSDSs	3000	32000	1288	867	30000
Chemical groupings stressed	—	—	—	NTP studied	DOD/GSA chemicals
Revision	quarterly	60 days	tertiary	—	quarterly
Information enhancement	yes	no	yes	yes	no
Readability	good	—	good	good	—
Access	good	good	good	good	—
Media	fiche online/tape paper	fiche	paper diskette fiche	paper	fiche online/tape
Cost (approx.) in dollars	3000+	4400+	325	270	140+
MSDS source main	literature	manufacturer	literature	NTP suppliers	DOD/GSA
Number of trade names	—	high	moderate	—	high

GE = General Electric, DOD = Department of Defense, GSA = General Services Administration, NTP = National Toxicology Program.

VCH Publishers, 303 NW 12th Avenue, Deerfield Beach, FL 33442-1788, (305) 428-5566

"Hazardous Materials Information Services" (HMIS), DOD 6050.5L, U.S. Government Printing Office, Washington, DC 20402

In order to find out the chemical components of products which do not list all ingredients on the label, you may have to request an MSDS from the manufacturer or supplier of the product. A sample form letter to help you request an MSDS is shown in Figure B.1.

[Date]

[Name]

[Address]

Dear Sir or Madam:

The New Jersey Worker and Community Right to Know Act (N.J.S.A. 34:5A-1 *et seq.*), effective August 29, 1984, establishes a comprehensive system for the disclosure and dissemination of information about hazardous substances in the workplace and the environment.

Pursuant to the Act, the Department of Health has adopted a Right to Know Hazardous Substance List (N.J.A.C. 8:59-9) which includes 2865 hazardous substances that pose a threat to the health and safety of employees. I am required to label the top five ingredients plus all hazardous substances down to 1.0%, and some down to 0.1%.

One or more of the products that we purchase from you do not contain a complete list of ingredients on the label. In order to comply with the Worker and Community Right to Know Act, we are requesting that you provide us with the Material Safety Data Sheet (MSDS) or the complete list of ingredients of the products indicated below and on the attached page(s).

Thank you for your assistance in this matter.

Sincerely,

Catalog No.   **Product Name**

_____   _____
_____   _____
_____   _____
_____   _____

Figure B.1.   Sample form letter to request an MSDS (developed by the New Jersey State Department of Health).

3. SARA TITLE III RESOURCES

National

For general information about the federal program or to receive materials, such as the law, regulations, fact sheets, etc., call the EPA Title III Hotline at (800) 535-0202.

New Jersey

For information related specifically to New Jersey's implementation of the federal program or reporting requirements, contact the New Jersey Department of Environmental Protection.

New Jersey Department of Environmental Protection
Division of Environmental Quality
Bureau of Hazardous Substances Information
401 East State Street, CN 405
Trenton, NJ 08625
(609) 292-6714

For specific questions about emergency planning or local emergency planning committees, contact:

New Jersey State Police
Office of Emergency Management
Box 7068
River Road
West Trenton, NJ 08628-0068
(609) 882-2000 (ask to be connected to Emergency Management)

4. SMOKING IN THE WORKPLACE RESOURCES

National

American Lung Association
1740 Broadway
New York, NY 10019-4374
(212) 315-8700

The national office of the American Lung Association will refer to local offices. Services at local offices include literature and audiovisuals on smoking, indoor air pollution, and lung disease. The smoking cessation program "Freedom From Smoking" can be offered at the worksite.

American Cancer Society
3340 Peachtree Road N.E.
Atlanta, GA 30326
(404) 320-3333

The American Cancer Society supplies, free of charge, literature, films, posters, speakers, and public education programs. The "FreshStart" smoking cessation program is available and can be held at the worksite. The national office will refer to local offices.

Office of Disease Prevention and Health Promotion
Health Information Center
P.O. Box 1133
Washington, DC 20013-1133
(800) 336-4797 or (202) 429-9091

The Office of Disease Prevention and Health Promotion is a clearinghouse for health information, providing publications on health and smoking, as well as the "Decision Maker's Guide."

Office of Technology Assessment
Health Program
United States Congress
Washington, DC 20510-8025
(202) 228-6590

The Health Program has published a staff paper entitled "Passive Smoking in the Workplace: Selected Issues," which describes effects of passive smoking and emphasizes workplace smoking policies.

The Tobacco Institute
1875 I Street, N.W.
Washington, DC 20006
(202) 457-4800

The Tobacco Institute supplies information on smoking in the workplace and sample policies that attempt to accommodate smokers and nonsmokers. The Institute produces a package of materials entitled "Workplace Smoking Restrictions: Some Considerations."

New Jersey

Smoking and Tobacco Use Control Program
New Jersey State Department of Health
CN 360
Trenton, NJ 08625
(609) 588-7470

The DOH issues written notices to employers upon receiving complaints from employees about compliance with the law, and they also provide consultation services to employers seeking to comply with the law. Also available from the DOH is a pamphlet entitled "Guidelines for Establishing a

Policy for Controlling Smoking in the Workplace," as well as copies of the law.

American Lung Association of New Jersey
2441 Route 22 West
Union, NJ 07083
(201) 687-9340

The American Lung Association (ALA) has available the booklets "Taking Executive Action" and "Creating Your Company Policy" which contain information and guidelines for establishing a smoking policy. The ALA offers the "Freedom From Smoking" smoking cessation/reduction program, and provides "No Smoking" signs and general information about smoking.
See Appendix D, Section 2 for New Jersey regional offices.

American Cancer Society
CN 2201
North Brunswick, NJ 08920
(201) 297-8000

The American Cancer Society (ACS) has "No Smoking" signs and information on smoking and cancer. They offer cancer education programs and the "FreshStart" smoking cessation program. ACS also has a "Model Policy for Smoking in the Workplace."
See Appendix D, Section 2 for local New Jersey offices.

New Jersey Group Against Smoking Pollution, Inc.
105 Mountain Avenue
Summit, NJ 07901
(201) 273-9368

The Group Against Smoking Pollution (GASP) provides consultation services to industries on regulating smoking. GASP also provides lists of smoke-free companies and restaurants, signs, videos, and general information, and has available a handbook entitled "Toward a Smoke-Free Workplace."

The Respiratory Health Association
55 Paramus Road
Paramus, NJ 07652
(201) 843-4111

The Respiratory Health Association provides advice and counsel on implementing corporate wellness programs, and can conduct health education classes at the workplace. It also has available smoking signs and the handbook, "Toward a Smoke-Free Workplace."

Environmental Improvement Associates
109 Chestnut Street
Salem, NJ 08079
(609) 935-4200

Environmental Improvement Associates is a nonprofit organization which supplies booklets on smoking policies for management and employees. In addition to supplying American Lung Association materials, the organization has available "Improving the Work Environment" for managers, and "Smoke-Free Work Areas" for employees.

New Jersey Chamber of Commerce Governmental Relations Office
240 West State Street, Suite 1518
Trenton, NJ 08608
(609) 989-7888

The New Jersey Chamber of Commerce maintains updates on what should be done for businesses concerning smoking legislation, supplying copies of the law and guidelines for policies.

New Jersey Business and Industry Association
102 West State Street
Trenton, NJ 08608
(609) 393-7707

The New Jersey Business and Industry Association will supply copies of the law, suggestions and recommendations for workplace policies, and "An Employer's Guide to the New Jersey Law Controlling Smoking in the Workplace."

APPENDIX C

Resources for Chapter Three

1. FEDERAL ENVIRONMENTAL LEGISLATION RESOURCES

General Publications

The Bureau of National Affairs, Inc.
9439 Key West Avenue
Rockville, MD 20850-3396
(800) 233-6067

Bureau of National Affairs (BNA) environment and safety publications include *Air Pollution Control, Chemical Regulation Reporter, Chemical Substances Control, Environment Reporter, Hazardous Materials Transportation, Index to Government Regulation, International Environment Reporter, International Hazardous Materials Transport Manual, Job Safety and Health, Loss Prevention and Control, Mine Safety and Health Reporter, Noise Regulation Reporter, Occupational Safety and Health Reporter, Product Safety and Liability Reporter, Sewage Treatment Construction Grants Manual, Toxics Law Reporter,* and *Water Pollution Control.*

Government Institutes, Inc.
966 Hungerford Drive, #24
Rockville, MD 20850
(301) 251-9250

Government Institutes (GI) has publications on environmental laws, RCRA, CERCLA/Superfund, waste minimization, hazardous materials, toxics, clean air, clean water, OSHA, international environment, information resources, cogeneration, energy management, and energy technology.

Lewis Publishers, Inc.
P.O. Drawer 519
121 South Main Street
Chelsea, MI 48118
(800) 525-7894

Lewis publishes hardcover reference and textbooks on Right-to-Know, emergency planning, occupational health, industrial hygiene, hazardous materials, toxicity and toxics, carcinogens, risk assessment, indoor and outdoor air contamination, air and water quality, hydrology, groundwater, drinking water and wastewater treatment, hazardous waste management, environmental law, health and health effects, pesticides and herbicides, toxicology, chemistry, industrial and municipal waste, hazardous and nonhazardous waste treatment and disposal, radiation, waste minimization, soils contamination, organic chemicals pollution, VOC, sludge technologies, and remedial technologies for leaking underground storage systems.

Pudvan Publishing Company
1935 Shermer Road
Northbrook, IL 60062
(312) 498-9840

Pollution Engineering is published monthly by Pudvan Publishing Company. Two annual issues contain an Environmental Control Telephone Directory. The May issue contains a directory for consultants, and the October issue contains a directory for manufacturers. These directories are available for purchase for $10 each.

General Information

EPA Small Business Hotline
(800) 368-5888

The EPA Small Business Hotline gives advice and information to small businesses on complying with EPA regulations and problems encountered by small quantity generators of hazardous waste and other small businesses with environmental concerns.

National Pesticide Telecommunications Network
(800) 858-7378

The Texas Technological University National Pesticide Telecommunications Network provides information on pesticide-related health/toxicity/minor cleanup to professionals and the general public.

Clean Water Act

National

See Section 3, "U.S. EPA Regional Offices."

New Jersey

New Jersey Department of Environmental Protection (DEP)
Division of Water Resources
Water Quality Management Element
Bureau of Permit Administration
401 E. State Street
CN 029
Trenton, NJ 08625
(609) 202-5262

Clean Air Act

National

See Section 3, "U.S. EPA Regional Offices."

New Jersey

New Jersey Department of Environmental Protection
Division of Environmental Quality
Bureau of New Source Review (Air Permits)
(609) 984-3032
Bureau of Enforcement Operations (Inspections and Investigations)
(609) 633-1129
401 E. State Street
CN 027
Trenton, NJ 08625

Toxic Substances Control Act

National

U.S. Environmental Protection Agency
Toxic Substances Control Assistance Office
401 M Street, SW
Washington, DC 20460
(202) 554-1404

As required by TSCA, EPA has established an office to provide technical and other nonfinancial assistance to chemical manufacturers, processors, and others who are interested in requirements and activities under this law. To help people understand TSCA's requirements, the TSCA Assistance Office (TAO) provides a telephone information service; a free bimonthly publication, *Chemicals and Progress Bulletin;* field consultants; and other technical assistance upon request. Also available is the free publication, *The Layman's Guide to the Toxic Substances Control Act.*

Publications

Government Institutes, Inc. publishes the *TSCA Handbook,* which covers TSCA regulations in detail. It can be purchased at the address listed above.

A major reference work identifying and classifying existing chemicals is published by the Office of Toxic Substances of the EPA. Known as the *Toxic Substances Control Act Chemical Substances Inventory,* it lists chemicals used for commercial purposes since 1975 as reported to EPA by domestic manufacturers and importers. The *Inventory* is an invaluable resource in determining if your chemical is a new or existing substance as defined by the Toxic Substance Control Act.

The most recent version, published in 1985, contains five volumes:

Volume I. CAS Registry Number

Volumes II & III. Substance Name Index

Volume IV. Molecular Form Index

Volume V. Unknown or Variable Composition Complex Reaction Products & Biological Material

The *Inventory* can be purchased for $161 for all five volumes (order #055-000-0025-4-1) from:

U.S. Government Printing Office
Room 110
26 Federal Plaza
New York, NY 10278
(212) 264-3825

The *Inventory* can also be found in various state libraries, although some may only have the initial *Inventory* from 1979.

The following New Jersey libraries have the *Inventory:*

Department of Environmental
  Protection Library
432 E. State Street, 1st Floor
Trenton, NJ 08625
(609) 984-2249

EPA Region II
GSA—Raritan Depot Building 209
Woodbridge Avenue
Edison, NJ 08837
(201) 321-6762

New Jersey State Library
185 W. State Street
Trenton, NJ 08625
(609) 292-6220

Rutgers University, Busch Campus
Library of Science and Medicine
Reference Area RA 1216, 2nd Floor
Bevier & Davidson Road

Piscataway, NJ 08854
(201) 932-2895

Rutgers University, Camden
  Campus
Camden Library
300 N. Fourth Street
Camden, NJ 08102
(609) 757-6034

Rutgers University, Newark Campus
Dana Library
185 University Avenue
Newark, NJ 07102
(201) 648-5901

Stevens Institute of Technology
S.C. Williams Library, 1st Floor
Castle Point
Hoboken, NJ 07030
(201) 420-5411

Resource Conservation and Recovery Act

National

U.S. Environmental Protection Agency
RCRA/Superfund Hotline
(800) 424-9346

Hotline services include interpretation of regulations and statutes for RCRA and Superfund, information on documentation, and referrals to agencies. The Hotline also provides information on the Underground Storage Tank Program.

New Jersey

New Jersey Department of Environmental Protection
Division of Hazardous Waste Management
Bureau of Regulation, Classification and Technical Assistance
401 E. State Street
CN 028
Trenton, NJ 08625
(609) 292-8341

CERCLA (Superfund)

National

See RCRA/Superfund Hotline, above

New Jersey

See "EPA Region II Directory," under Section 3, "U.S. EPA Regional Offices."

SARA Title III

Publications

Thompson Publishing Group publishes the *Community Right-to-Know Manual: The Guide to SARA Title III*. It can be purchased from:

Thompson Publishing Group
Subscription Service Center
P.O. Box 76927
Washington, DC 20013
(800) 424-2959

Lowrys' Handbook of Right-to-Know and Emergency Planning can be purchased from Lewis Publishers, Inc. (See "General Publications," above.)

Audiovisual Resources

"The Toxics Release Inventory: Meeting the Challenge" is a 19-minute overview videotape designed to explain toxic release reporting to plant facility managers and others who need to know about the requirement. It can be purchased from:

Color Film Corporation
Video Division
770 Connecticut Avenue
Norwalk, CT 06854
(800) 882-1120

National

Call the EPA Title III Hotline at (800) 535-0202. EPA will supply information on the federal regulations and State Emergency Response Commissions, and provide materials pertaining to the law.

New Jersey

For questions about the federal program or reporting requirements, contact:

New Jersey Department of Environmental Protection
Division of Environmental Quality
Bureau of Hazardous Substances Information
401 E. State Street
CN 405
Trenton, NJ 08625
(609) 292-6714

For specific questions about emergency planning or local emergency planning committees, contact:

New Jersey State Police
Office of Emergency Management
Box 7068
River Road
West Trenton, NJ 08628-0068
(609) 882-2000 (ask to be connected to Emergency Management)

2. NEW JERSEY ENVIRONMENTAL LEGISLATION RESOURCES

Toxic Catastrophe Prevention Act

New Jersey Department of Environmental Protection
Division of Environmental Quality
Bureau of Release Prevention
401 E. State Street
CN 027
Trenton, NJ 08625
(609) 633-7289

Spill Compensation and Control Act

New Jersey Department of Environmental Protection
Division of Water Resources
Bureau of Industrial Waste Management
401 E. State Street
CN 029
Trenton, NJ 08625
(609) 292-4860

To report a spill: New Jersey Department of Environmental Protection Hotline: (609) 292-7172

Environmental Cleanup Responsibility Act

New Jersey Department of Environmental Protection
Division of Waste Management
Industrial Site Evaluation Element
401 E. State Street
CN 028
Trenton, NJ 08625
(609) 633-7141

New Jersey Department of Environmental Protection Hazardous Waste Advisement Program

The Hazardous Waste Advisement Program is confidential, and independent of New Jersey Department of Environmental Protection's enforcement efforts. It assists hazardous waste generators in understanding and complying with New Jersey's hazardous waste regulations, which for small quantity generators are more stringent than federal regulations. The program distributes copies of regulations, fact sheets, and a newsletter. There also is an Information Hotline (see number below).

New Jersey Department of Environmental Protection
Division of Hazardous Waste Management
Hazardous Waste Advisement Program
401 E. State Street, 5th Floor
Trenton, NJ 08625
(609) 292-8341

3. U.S. EPA REGIONAL OFFICES

For a list of states by standard federal regions, see Appendix E, Section 1.

Region I

JFK Federal Building
Boston, MA 02203
(617) 565-3715

Region II

26 Federal Plaza
New York, NY 10278
(212) 264-2515

Field Component
Caribbean Field Office
P.O. Box 792
San Juan, PR 00909
(809) 725-7825

Region III

841 Chestnut Street
Philadelphia, PA 19107
(215) 597-9800

Region IV

345 Courtland Street NE
Atlanta, GA 30365
(404) 347-4727

Region V

230 South Dearborn Street
Chicago, IL 60604
(312) 353-2000

Field Component
Eastern District Office
25089 Center Ridge Road
West Lake, OH 44145
(216) 835-5200

Region VI

1445 Ross Avenue
Dallas, TX 75202
(214) 655-2200

Region VII

726 Minnesota Avenue
Kansas City, KS 66101
(913) 236-2800

Region VIII

One Denver Place
999 18th Street
Denver, CO 80202-2413
(303) 293-1603

Region IX

215 Fremont Street
San Francisco, CA 94105
(415) 974-8071

Field Component
Pacific Islands Office
P.O. Box 50003
300 Ala Moana Boulevard
Room 1302
Honolulu, HI 96850
(800) 546-8910

Region X

1200 Sixth Avenue
Seattle, WA 98101
(206) 442-5810

Field Components
Alaska Operations Office

Room E556, Federal Building
701 C Street
Anchorage, AK 99513
(907) 271-5083

Alaska Operations Office
3200 Hospital Drive

Juneau, AK 99801
(907) 586-7619

Idaho Operations Office
422 West Washington Street
Boise, ID 83702
(208) 334-1450

EPA Region II Directory

New Jersey, New York, Puerto Rico, Virgin Islands

26 Federal Plaza
New York, NY 10278
(Hours: 8:00 a.m.–6:00 p.m.)

Office	Phone
Regional Administrator	(212) 264-2525
Deputy Regional Administrator	(212) 264-0396
Executive Assistant	(212) 264-2525
Office of Regional Counsel	(212) 264-1017
Deputy Regional Counsel	(212) 264-1018
Associate Regional Counsel	(212) 264-4430
NY Superfund Branch	(212) 264-9858
NJ Superfund Branch	(212) 264-4940
Air Branch	(212) 264-8181
Waste and Toxic Substances Branch	(212) 264-5340
Water, Grants and General Law Branch	(212) 264-9885
Office of External Programs	
Director	(212) 264-2515
Deputy Director	(212) 264-4535
Office of Policy and Management	
Asst. Regional Administrator for Policy and Mgmt.	(212) 264-2520
Deputy Director	(212) 264-0455
Equal Employment Opportunity Officer	(212) 264-1709
Financial Management Branch	(212) 264-8989
Personnel and Organization Branch	(212) 264-0016
Facilities and Administrative Management Branch	(212) 264-1414
Information Systems Branch	(212) 264-9850
Grants Administration Branch	(212) 264-9860
Environmental Impacts Branch	(212) 264-1892
Permits Administration Branch	(212) 264-9881
Policy and Program Integration Branch	(212) 264-4296
Planning and Evaluation Branch	(212) 264-3052

EPA Region II Directory, continued

Office	Phone
Air and Waste Management Division	
Director	(212) 264-2301
Deputy Director	(212) 264-3082
Air Programs Branch	(212) 264-2517
Hazardous Waste Program Branch	(212) 264-7309
Hazardous Waste Compliance Branch	(212) 264-6151
Hazardous Waste Facilities Branch	(212) 264-0505
Air Compliance Branch	(212) 264-9627
Radiation Representative	(212) 264-4418
Water Management Division	
Director	(212) 264-2513
Deputy Director	(212) 264-2514
Groundwater Coordinator	(212) 264-5635
New Jersey Municipal Programs Branch	(212) 264-5692
New York Municipal Programs Branch	(212) 264-0959
Water Permits and Compliance Branch	(212) 264-9894
Marine and Wetlands Protection Branch	(212) 264-5170
Water Standards and Planning Branch	(212) 264-1833
Drinking/Groundwater Protection Branch	(212) 264-1800
Emergency and Remedial Response Division	
Director	(212) 264-8672
Deputy Director	(212) 264-1574
Program Support Branch	(212) 264-3984
Site Compliance Branch	(212) 264-2647
NJ Remedial Action Branch	(212) 264-1872
NY Remedial Action Branch	(212) 264-0106
Response and Prevention Branch	(201) 321-6658
Environmental Services Division (Edison, NJ)	
Director	(201) 321-6754
Deputy Director	(201) 321-6755
Surveillance and Monitoring Branch	(201) 321-6686
Monitoring Management Branch	(201) 321-6645
Pesticides and Toxic Substances Branch	(201) 321-6765
Technical Support Branch	(201) 321-6706

4. NEW JERSEY DEP DIRECTORY

DEP Hotline: To report abuses of the environment, call (609) 292-7172, 24 hours a day.

Office	Phone
Air Pollution Report & Forecast (recording)	(609) 392-1463
Coastal Resources	(609) 292-2795
Communications	(609) 292-2103
Emergency Response (24-hour hotline)	(609) 292-7172
Enforcement	
Air/Noise/Radiation/Pesticide	(609) 633-7288

New Jersey DEP Directory, continued

Office	Phone
Coastal Resources	(609) 292-5120
Hazardous Waste	(609) 633-0700
Solid Waste	(609) 426-0700
Water Resources	(609) 426-0799
Environmental Cleanup Responsibility Act	(609) 633-7141
Environmental Education & Awareness	(609) 984-7478
Environmental Impact Statements	(609) 292-2662
Environmental News	(609) 984-6773
Environmental Quality	(609) 292-5383
Air Pollution	(609) 292 6704
Emergency Response	(609) 530-4031
Accidental Release Prevention	(609) 633-7289
Right to Know	(609) 292 6714
Laboratory	(609) 530-4119
Noise Control	(609) 984-4161
Pesticide Control	(609) 530-4123
Radiation Protection	(609) 530-4001
Financial Management & General Services	(609) 292-9230
Fish/Game/Wildlife	(609) 292-2965
Green Acres	(609) 588-3450
Harbor Cleanup	(609) 292-5990
Hazardous Site Mitigation	(609) 984-2902
Hazardous Waste Management	(609) 292-1250
New Jersey Outdoors	(609) 292-2477
Parks & Forestry	(609) 292-2733
Permits	
Air Quality	(609) 984-3032
Coastal (CAFRA, Riparian, Wetlands)	(609) 292-0060
Hazardous & Solid Waste	(609) 292-5196
Pesticide Certification	(609) 530-4133
Water Permit Information	
NJPDES (Discharge)	
Industrial	(609) 292-0407
Municipal	(609) 984-4429
Groundwater	(609) 292-0424
Sewer Connections	(609) 984-4429
Stream Encroachment	(609) 292-2373
Water Quality Certificates	(609) 633-7026
Personnel & Data Processing	(609) 292-1898
Planning Group	(609) 292-2662
Regulatory & Governmental Affairs	(609) 292-9320
Resources Interpretive Service	(609) 633-2103
Right to Know	(609) 633-7289
Science & Research	(609) 984-6070
Solid Waste Management	(609) 292-8879
Recycling	(609) 292-9450
Resource Recovery	(609) 292-8503
Solid Waste Planning	(609) 292-8242
Water Resources	(609) 292-1637
Construction Grants	(609) 292-8961
Geological Survey	(609) 292-1185
Water Quality Management	(609) 292-5262
Water Quality Monitoring	(609) 633-7010
Water Supply	(609) 292-7219

Resources for Chapter Four

For additional resources in Industrial Hygiene, see Appendix A.

1. EMPLOYEE EDUCATION AND TRAINING RESOURCES

Suggested Readings in Industrial Safety and Health Training

The following publications give guidelines for developing education/training programs:

"Training Requirements in OSHA Standards and Training Guidelines" OSHA 2254, OSHA Publications Distribution Office, U.S. Department of Labor, Room N-3101, 200 Constitution Avenue, N.W., Washington, DC 20210

"OSHA Publications and Audiovisual Programs" OSHA 2019, OSHA Publications Distribution Office, U.S. Department of Labor, Room N-3101, 200 Constitution Avenue, N.W., Washington, DC 20210

"OSHA Safety and Health Training Guidelines for General Industry" (PB-239-310/AS), National Technical Information Service, Springfield, VA 22161

"OSHA Safety and Health Training Guidelines for Maritime Operations" (PB-239-311/AS), National Technical Information Service, Springfield, VA 22161

"OSHA Safety and Health Training Guidelines for Construction" (PB-239-312/AS), National Technical Information Service, Springfield, VA 22161

"Worker and Community Right to Know Act, Education and Training

Program Guide," 1986, Right to Know Project, New Jersey State Department of Health, CN 368, Trenton, NJ 08625-0368

Supervisor's Safety Manual, 1988, 9th Edition, National Safety Council, Orders Dept., 444 North Michigan Avenue, Chicago, IL 60660

"Media Resource Catalog," National Audiovisual Center, 8700 Edgeworth Drive, Capitol Heights, MD 20743-3701

Companies Producing Materials for Employee Education and Training

Filmedia, Inc.
10740 Lyndale Avenue South
Minneapolis, MN 55420
(800) 328-3789

International Film Bureau, Inc.
332 S. Michigan Avenue
Chicago, IL 60604-4382
(312) 427-4545

McGraw-Hill Films
P.O. Box 641
Del Mar, CA 92014
(800) 548-0660

Tel-A-Train, Inc.
P.O. Box 410
Halls Road
Old Lyme, CT 06371
(800) 225-2026

BNA Communications, Inc.
9439 Key West Avenue
Rockville, MD 20850-3396
(800) 233-6067

National Audio Visual Center
National Archives and
Records Service
8700 Edgeworth Drive
Capitol Heights, MD 20743-701
(800) 638-1300

DuPont Safety Services
Barley Mill P19-1210
Wilmington, DE 19898
(800) 532-SAFE

UNZ & Co.
190 Baldwin Avenue
Jersey City, NJ 07306
(201) 795-5400
(800) 631-3098

Business and Legal Reports
64 Wall Street
Madison, CT 06443
(800) 553-4569

Fire Prevention Through Films, Inc.
P.O. Box 11
Newton Highlands, MA 02161
(617) 965-4444

Industrial Training Systems Corp.
20 West Stow Road
Marlton, NJ 08053-9990
(609) 983-7300
(800) 922-0782

Mine Safety Appliances, Inc.
1100 Globe Avenue
Mountainside, NJ 07092
(800) MSA-2222

Coronet MPI
Film Communicators, Inc.
108 Wilmont Road
Deerfield, IL 60015
(800) 621-2131

National Society for the Prevention
   of Blindness
500 E. Remington Road
Schaumburg, IL 60173
(312) 843-2020

Visucom Productions, Inc.
P.O. Box 5472
Redwood City, CA 94063
(800) 222-4002

Gulf Publishing Company Video
P.O. Box 2608
Houston, TX 77252
(713) 529-4301

Industrial Training Systems, Inc.
20 West Stow Road
Marlton, NJ 08053-9990
(609) 983-7300
(800) 922-0782
(produces some materials in French
   and Spanish)

Federal Agencies

The following federal governmental agencies can provide information and educational materials on general occupational safety and health matters.

Occupational Safety and Health Administration (OSHA), United States Department of Labor

(For a listing of the national and regional OSHA offices, see Appendix B, Section 1.) OSHA publishes a variety of pamphlets and booklets on specific toxic substances and other occupational health and safety hazards regulated by the federal government. These materials are listed in the booklet *OSHA Publications and Audiovisual Programs* (OSHA 2019, 1986 revised). OSHA publications may be ordered from regional OSHA offices.

Selected OSHA publications include:

"A Guide to Worker Education Materials in Occupational Safety and Health" (1982), a listing of materials produced by private and nonprofit organizations.

"All About OSHA," OSHA 2056 (1985 revised), a good brief summary of the OSHA law, coverage, standards, and penalties.

"OSHA Handbook for Small Businesses," OSHA 2209 (1979 revised), a handbook to assist small business employers in meeting their legal requirements under the OSHAct.

Code of Federal Regulations, Title 29, Parts 1900–1910. This volume, parts 1900–1910, represents all current regulations codified under Title 29 (Labor) as of July 1. (Volumes are revised yearly.)

"Personal Protective Equipment," OSHA 3077 (1985), discusses types of equipment most commonly used for protection of the head, torso, arms and hands, and feet.

OSHA Training Institute. The OSHA Training Institute encompasses a wide range of training using laboratories, seminars, models, and demonstrations. The Institute is open to safety and health officers from federal, state, and local governments, as well as to representatives from private industry. More than 60 different courses are offered, and OSHA is constantly upgrading and developing courses as new technology demands, as new needs arise, and as industrial safety and health concerns warrant.

U.S. Department of Labor
OSHA Training Institute
1555 Times Drive
Des Plaines, IL 60018
(312) 297-4810

National Institute for Occupational Safety and Health (NIOSH)

(For a listing of the national and regional offices, see Appendix B, Section 1.) The principal federal agency involved in research, education, and training in occupational health, NIOSH provides detailed information and written materials on health hazards. NIOSH has a listing of videotapes for loan (free of charge), rental, and sale. Publications relating to occupational health hazards may be obtained by contacting the Publications Department. The 1987 NIOSH Publications Catalog is also available.

NIOSH also offers training courses in occupational safety and health at training centers throughout the country. For information, contact the Division of Training and Manpower.

NIOSH funds 14 Educational Resource Centers at universities throughout the country, providing continuing education programs through short courses and workshops in the fields of occupational safety, industrial hygiene, occupational medicine, and occupational health nursing.

NIOSH Educational Resource Centers

Alabama Educational Resource
  Center
Deep South Center for Occupational
  Health and Safety
University of Alabama at
  Birmingham
School of Public Health

Medical Center, University Station
Birmingham, AL 35294
(205) 934-7032

Northern California Educational
  Resource Center
School of Public Health

University of California at Berkeley
206 Earl Warren Hall
Berkeley, CA 94720
(415) 642-0761

Southern California Educational
    Resource Center
Institute of Safety and Systems
    Management
Center for Occupational Safety and
    Health
University of Southern California
Los Angeles, CA 90089-0021
(213) 743-8983

Cincinnati Educational Resource
    Center
Institute of Environmental Health
University of Cincinnati
3223 Eden Avenue
Cincinnati, OH 45267
(513) 558-5701

Harvard Educational Resource
    Center
Department of Environmental
    Science and Physiology
Harvard School of Public Health
665 Huntington Avenue
Boston, MA 02115
(617) 732-1260

Illinois Educational Resource Center
University of Illinois at Chicago
Occupational Health and Safety
    Center
817 South Wolcott Avenue
Chicago, IL 60612
(312) 996-7887

The Johns Hopkins Educational
    Resource Center
School of Hygiene and Public Health

615 North Wolfe Street
Baltimore, MD 21205
(301) 955-3602

Michigan Educational Resource
    Center
Center for Occupational Health and
    Safety Engineering
The University of Michigan
1205 Beal Avenue, IOE Building
Ann Arbor, MI 48109-2117
(313) 763-2245

Minnesota Educational Resource
    Center
Midwest Center for Occupational
    Health and Safety
School of Public Health
University of Minnesota
1162 Mayo Memorial Building
420 Delaware Street, S.E.
Minneapolis, MN 55455
(612) 221-8770

New York/New Jersey Educational
    Resource Center
Universities Occupational Safety and
    Health Educational Resource
    Center
Mt. Sinai School of Medicine
1 Gustave Levy Place
New York, NY 10029
(212) 241-4804

North Carolina Educational
    Resource Center
School of Public Health
University of North Carolina
109 Conner Drive, 346A, Suite 1101
Professional Village
Chapel Hill, NC 27514
(919) 962-2101

Texas Educational Resource Center
The University of Texas Health
  Science Center at Houston
School of Public Health
P.O. Box 20186
Houston, TX 77225-0186
(713) 792-4638

Utah Educational Resource Center
Rocky Mountain Center for
  Occupational and Environmental
  Health
University of Utah Medical Center
Department of Family and
  Community Medicine, Room BC
  106

Salt Lake City, UT 84132
(801) 581-8719

Washington Educational Resource
  Center
Northwest Center for Occupational
  Health and Safety
Department of Environmental
  Health, SC-34
University of Washington
Seattle, WA 98195
(206) 543-6991

Organizations

The network of organizations providing educational materials on occupational health and safety is extensive. The following organizations can either inform employers and employees of available materials or provide appropriate materials. This list was adapted from the New Jersey State Department of Health, Occupational Disease Prevention and Information Program, Right to Know Resources.

National and New Jersey Organizations

American Society of Safety Engineers (ASSE)
1800 East Oakton Street
Des Plaines, IL 60018-2187
(312) 692-4121

ASSE provides publications, course materials, audiovisual packages, safety films, and seminars designed for employees, safety personnel, supervisors, or management. Call or write for resource brochure.

National Safety Council
444 North Michigan Avenue
Chicago, IL 60660
(312) 527-4800

The National Safety Council (NSC) publishes a national directory of safety films and a catalog and poster directory containing publications on occupational safety and health. The New Jersey Chapter conducts occupational safety and health and Right to Know seminars and workshops. In New Jersey, the NSC is located at 6 Commerce Drive, Cranford, NJ 07016, (201) 272-7712.

American Lung Association
1740 Broadway
New York, NY 10019-4374
(212) 315-8700

The national office of the American Lung Association (ALA) will refer to local offices. ALA publishes a booklet, "Occupational Lung Disease—An Introduction," detailing lung disease and its major occupational causes.

Center for Occupational Hazards
5 Beekman Street
New York, NY 10038
(212) 227-6220

The Center for Occupational Hazards (COH), a national clearinghouse for research and education in hazards in the arts, publishes *Art Hazards Newsletter,* maintains an art hazards information center, and sponsors courses for artists, schoolteachers, and others in occupational health.

Women's Occupational Health Resource Center
117 St. John's Place
Brooklyn, NY 11217
(718) 230-8822

The Women's Occupational Health Resource Center (WOHRC) publishes a newsletter and produces fact sheets and packets on the hazards in occupations dominated by women.

Environmental Hazards Management Institute (EHMI)
P.O. Box 283
137 High Street
Portsmouth, NH 03801
(603) 436-3950

The Environmental Hazards Management Institute (EHMI) provides several services regarding hazardous materials/waste management, including sponsoring the annual HazMat Conference (in Atlantic City in June); confidential environmental audits; employer and employee training; executive briefings on laws and liabilities; management of corrective action cleanup activities; training films; and trainings in occupational safety and health, RCRA, and OSHA Regulation 1910.120.

Unions

Many international and local unions have produced health and safety materials and held programs to educate workers on the hazards of specific occupations. Union newsletters frequently highlight occupational health hazards and their prevention. See OSHA's "A Guide to Worker Education Materials in Occupational Safety and Health" for details. (See "Selected OSHA Publications," under "Federal Agencies," above.) The following are examples of organizations providing these services:

International Molders and Allied Workers Union
Health and Safety Department
1225 East McMillan Street, Suite 302
Cincinnati, OH 45206
(513) 221-1526

The International Molders and Allied Workers Union (IMAWU) publishes *Health Hazards: The Ignored Reality* ($10), an overview of hazard recognition and control, and *Health and Safety Resource Directory* ($5), a compilation of occupational health and safety resource groups.

International Brotherhood of Painters and Allied Trades
1750 New York Ave., N.W.
Washington, DC 20006
(202) 637-0700

The International Brotherhood of Painters and Allied Trades (IBPAT) provides training and education for members on paint and allied products, their safe use, diseases and illnesses associated with them, and methods of identification and protection. Catalogs for printed and audiovisual materials can be obtained by contacting IBPAT.

Worker's Institute for Safety and Health
1126 16th Street, N.W.
Washington, DC 20036
(202) 887-1980

The Worker's Institute for Safety and Health (WISH) conducts specific programs on occupational safety and health at the request of both union and employer.

Committees on Occupational Safety and Health (COSH)

COSH groups are independent local organizations of workers, local unions, and health, safety, and legal professionals concerned with industrial health hazards and worker protection. They provide information to employees and unions. (See Appendix A for a listing of COSH groups.)

University Programs on Occupational Safety and Health

Occupational Safety and Health Education Center
Institute of Management and Labor Relations
Rutgers, The State University of New Jersey
Ryders Lane
New Brunswick, NJ 08903
(201) 932-9242

The Occupational Safety and Health Education Center of the Institute of Management and Labor Relations at Rutgers, The State University of New Jersey offers a variety of educational services and programs for employers and other agencies.

Public Education and Risk Communication Division
Environmental and Occupational Health Sciences Institute
UMDNJ–Robert Wood Johnson Medical School and Rutgers, the State University
    of New Jersey
675 Hoes Lane, TR 2
Piscataway, NJ 08854-5635
(201) 463-4500

The Public Education and Risk Communication Division of the Environmental and Occupational Health Sciences Institute (EOHSI), which is jointly administered by the University of Medicine and Dentistry of New Jersey (UMDNJ)–Robert Wood Johnson Medical School and Rutgers, The State University of New Jersey, publishes the "Directory of Organizations Involved in Environmental and Occupational Health." Other parts of this division are the Mid-Atlantic Asbestos Training Center supported by EPA, the Continuing Education and Outreach Program of the University's Occupational Safety and Health Educational Resource Center (see "NIOSH Educational Resource Centers," under "Federal Agencies," above), and the New York/New Jersey Hazardous Materials Worker Training Center supported by the National Institute of Environmental Health Services (NIEHS).

The Management Institute
School of Business Administration
Glassboro State College

Glassboro, NJ 08028
(609) 863-5392

The Management Institute, in association with Glassboro State College, provides formal management training for the business and industry community. The program comprises seminars, workshops, courses, meetings and in-house training, and includes a course on supervising workers who handle hazardous materials.

Ohio State University
Department of Photography and Cinema
156 West 19th Avenue
Columbus, Ohio 43210-1183
(614) 292-5966

The university's Labor Education and Research Service developed the Hazard Recognition Training Program under a contract from OSHA. The program's audiovisual materials consist of 18 units (26 films) that can be used to build a wide variety of training programs and courses in the areas of construction safety, industrial safety, and occupational health. Call for film catalog.

Insurance Companies, Industrial Supply Companies, Consultants, etc.

Many of these organizations are able to provide businesses with general assistance on occupational safety and health and the toxicity of specific chemicals.

Right to Know Project
New Jersey State Department of Health
CN 368
Trenton, NJ 08625-0368
(609) 984-2202

The Right to Know Project of the New Jersey State Department of Health publishes a resource list of consultants who offer training and assistance on Right to Know. Many of these companies also offer various occupational health training programs.

3M Company
Occupational Safety and Health Division
3M Center
220-7W-02
St. Paul, MN 55144
(800) 243-4630 (technical information), (800) 328-1667 (literature, local representatives)

3M manufactures a full line of respiratory protection devices. They have information on respirator selection, fitting, use, and training. Call for catalog.

Producers of Educational Materials

Many companies produce educational posters, literature, and audiovisual materials on occupational safety and health topics for employees. A listing of these companies is provided near the beginning of this Appendix. See "Organizations" (Section 2) for information on companies providing materials for employee health promotion.

Occupational Health and Safety
Medical Publications, Inc.
225 N. New Road
Waco, TX 76710
(817) 776-9000

This company publishes the *1988/89 Occupational Health and Safety Purchasing Sourcebook*, a comprehensive guide for employers. The directory includes products and services, company profiles, trade/brand names, and toll-free numbers ($29.95).

New Jersey Agencies

The New Jersey State Department of Health. DOH provides occupational safety and health information and materials relating to private sector workplaces. Staff industrial hygienists will perform workplace inspections in response to employee and employer requests. DOH also conducts NIOSH-assigned Health Hazard Evaluations in the state of New Jersey.

New Jersey State Department of Health
Occupational Health Service
CN 360
Trenton, NJ 08625-0360
(609) 984-1863

Occupational Safety and Health Consultation Services. This agency performs plant inspections and helps industries to design methods of hazard control and means of meeting federal OSHA requirements. For more detailed information, see Chapter 5, Section 2.

New Jersey Department of Labor
OSHA Consultation Services
CN 054
Trenton, NJ 08625-0054
(609) 984-3507

Public Employee Occupational Safety and Health Project. Through the Office of Public Employees Safety, this agency provides occupational health information, education, and training materials to public employees and employers and inspects public workplaces for violations of Public Employees Occupational Safety and Health Act (PEOSHA) health standards.

New Jersey Department of Labor
Office of Public Employees Safety
Station Plaza Four, 3rd Floor
CN 386
28 Yard Avenue
Trenton, NJ 08625
(609) 292-7036

Trade Associations. Business groups may be able to assist member businesses with occupational health and safety and New Jersey Right to Know information.

New Jersey State Chamber of Commerce (NJSCC)

Five Commerce Street
Newark, NJ 07102
(201) 623-7070

315 West State Street
Trenton, NJ 08609
(609) 989-7888

The NJSCC has a Right to Know Compliance Service and conducts seminars and workshops on Right to Know and occupational health.

The Chemical Industry Council of New Jersey (CIC)
150 West State Street
Trenton, NJ 08609
(609) 392-4214

CIC conducts seminars for chemical manufacturing employers and employees on various legislative and regulatory initiatives including occupational health issues.

New Jersey Area OSHA Offices

Hunterdon, Middlesex, Somerset,
Union, and Warren Counties

Avenel Area OSHA Office
  Plaza 35, Suite 205
1030 St. Georges Avenue
Avenel, NJ 07001
(201) 750-3270

Atlantic, Burlington, Camden,
Cape May, Cumberland,
Gloucester, Mercer, Monmouth,
Ocean, and Salem Counties

Camden Area OSHA Office
2101 Ferry Ave., Room 403
Camden, NJ 08104
(609) 757-5181

Essex, Hudson, Morris, and
Sussex Counties

Dover Area OSHA Office
2 Blackwell Street
Dover, NJ 07801
(201) 361-4050

Bergen and Passaic Counties

Hasbrouck Heights Area Office
Teterboro Airport Professional
  Building
500 Route 17
Hasbrouck Heights, NJ 07604
(201) 288-1700

New Jersey Local/County Occupational Health Programs

City of Elizabeth
City Hall
50 Winnfield Scott Plaza
Elizabeth, NJ 07201
(201) 820-4049 or 4060
(covers Elizabeth and Union County)

Occupational Health Care
  Consortium of Northern NJ
Paterson Health Department
176 Broadway
Paterson, NJ 07505
(201) 881-3914
(covers Passaic County)

Hudson Regional Health
  Commission
215 Harrison Avenue
Harrison, NJ 07029

(201) 485-7001
(covers Hudson County)

Atlantic County Department of
  Health
201 South Shore Road
Northfield, NJ 08225
(609) 645-7700 ext. 4372
(covers Atlantic County)

Bergen County Department of
  Health Services
327 Ridgewood Avenue
Paramus, NJ 07652-4895
(201) 599-6100
(covers Bergen County)

Clifton Health Department
900 Clifton Avenue
Clifton, NJ 07011
(201) 470-5758
(covers City of Clifton)

Atlantic City Division of Health
35 South Illinois Avenue
Atlantic City, NJ 08401
(609) 347-5671
(covers Atlantic City)

Note: At the time of this writing, the occupational health programs throughout the State of New Jersey are few. However, New Jersey requires all Health Departments to have occupational health programs by January 1989.

2. EMPLOYEE HEALTH PROMOTION RESOURCES

Selected Publications

The Office of Disease Prevention and Health Promotion
National Health Information Center
U.S. Department of Health and Human Services
P.O. Box 1133
Washington, DC 20013-1133
(800) 336-4797

Available from this source are the following publications:
"Small Business and Health Promotion: The Prospects Look Good. A Guide for Providers of Health Promotion Programs" (1984). Single copy available, Order No. W0004, $2 handling fee.
"Worksite Health Promotion: A Bibliography of Selected Books and Resources" (1985). Single copy available, Order No. W0005, $2 handling fee.
"Worksite Nutrition: A Decision Maker's Guide" (1986). Single copy available, Order No. U0010, $2 handling fee.

The Washington Business Group on Health
Worksite Wellness Series
229$^1/_2$ Pennsylvania Avenue, S.E.
Washington, DC 20003
(202) 547-6644

This group publishes the Worksite Wellness Media Reports (includes "Wellness in Small Businesses"). Order form listing titles and scheduled dates of publication is available.

The President's Committee on Employment of the Handicapped, 1111 20th Street NW, Washington, D.C. 20036, (202) 653-5044.

"The Employer's Guide, How to Successfully Supervise Employees with Disabilities" can be obtained free of charge.

Nonprofit Organizations: National and New Jersey

Various resources are available either free of charge or at minimal cost. These most often are offered through nonprofit agencies in the community. Literature, films, speakers, or classes are available through volunteer organizations. Local health departments or hospitals may have health educators or public health nurses who can conduct classes or health screenings.

The Cancer Information Service of the National Cancer Institute provides the latest information on the causes, prevention, and detection of cancer, and has available free pamphlets on a wide variety of topics, written in lay language for cancer patients, their families, and the general public. There are 25 local Cancer Information Service offices throughout the United States. The toll-free number for the national office is (800) 4-CANCER.

The American Cancer Society (ACS) supplies literature, films, posters, speakers, and public education programs, free of charge. The "FreshStart" smoking cessation program is available for a fee, and can be held at the worksite. (See "National and New Jersey Local Offices," below.) Also available is the "Taking Control" program, which lists ten steps to a healthier life and reduced risk of cancer.

The American Heart Association (AHA) provides literature, posters, films, and speakers on preventing heart disease. The materials are available in large quantities at low cost. "Heart at Work" is a worksite health promotion program offered in components. The employer has a choice of which components may be used, and to what degree the employees may be involved. "Heart at Work" is available at the state AHA office. Other AHA materials are available through the regional offices. (See "National and New Jersey Local Offices.")

The American Lung Association (ALA) supplies literature (in limited quantities) and audiovisuals on smoking, indoor air pollution, and lung disease, free of charge. There is a fee for large quantities of literature and for ALA's smoking cessation/reduction program, "Freedom From Smoking," at the worksite (See "National and New Jersey Local Offices.")

Other volunteer organizations include local chapters of the American Diabetes Association, or the New Jersey state office of the Arthritis Foundation, at 15 Prospect Lane, Colonia, NJ 07067, (201) 388-0744.

Many County Cooperative Extension offices provide worksite programs on nutrition and weight control, have lending libraries for audiovisuals, and can supply limited materials free or at low cost.

The March of Dimes has a free program called "Good Health is Good Business." The program can be tailored to a company's needs and includes information on health habits and childbearing, early prenatal care, genetic counseling, and communicating with children. Contact the March of Dimes Birth Defects Foundation, Community Services, 1275 Mamaroneck Avenue, White Plains, NY 10605, (914) 428-7100.

Special interest groups such as the American Red Cross or YM/YWCAs also may have courses, teachers, or materials available. Y's usually have gym facilities and pools.

Many hospitals are implementing lifestyle, occupational health, and employee assistance programs for local businesses and industries. They may also provide onsite health screenings for employees.

Public health departments on the local and county levels often have health educators who will provide onsite programs free of charge. Public health nurses may be available to do onsite health screenings such as blood pressure tests.

Other organizations that may supply materials or programs or refer you to resources would be local medical, dental, or hospital associations, community organizations such as United Way, or civic groups such as the Lions Club. The American Medical Association (AMA) provides literature for the general public on health promotion topics. Call AMA Publications at (800) 621-8335.

There are some companies that produce educational materials (posters, literature, and audiovisuals) on occupational health and health promotion topics. These items can be purchased or often rented. A product catalog is available from Krames Communications, 312 90th Street, Daly City, CA 94015-1898, (800) 228-8347. Educational booklets by Scriptographic on safety and health topics are available from Channing L. Bete Co., Inc., 200 State Road, South Deerfield, MA 01373, (800) 628-7733. Videotapes on health, nutrition, and fitness are available from RMI Media Productions, Inc., 2807 West 47th Street, Shawnee Mission, KS 66205, (800) 821-5480. (Also see "Companies Producing Materials for Employee Education and Training" in Section 1 of this appendix.)

Local video stores sell or rent tapes on health or exercise. Municipal fire and police departments often provide lectures on fire prevention and personal safety.

American Cancer Society Offices

National Office:

American Cancer Society
3340 Peachtree Road N.E.
Atlanta, GA 30326
(404) 320-3333

New Jersey State Office:

New Jersey Division, Inc.
2600 Route 1
CN 2201
North Brunswick, NJ 08902
(201) 297-8000

New Jersey County Offices

Atlantic
101 S. Shore Road
Northfield, NJ 08225
(609) 645-7272

Bergen
20 Mercer Street
Hackensack, NJ 07601
(201) 343-2222

Burlington
P.O. Box 1160
Woodlane Road
Mt. Holly, NJ 08060
(609) 267-8444

Camden
410 White Horse Pike
Haddon Heights, NJ 08035
(609) 546-1600

Cape May
15 Delsea Drive
Rio Grande, NJ 08242
(609) 886-1154

Cumberland
1400 W. Landis Avenue
Vineland, NJ 08360
(609) 692-1364

Essex
336 S. Harrison Street

E. Orange, NJ 07018
(201) 678-1990

Gloucester
110 E. High Street
Glassboro, NJ 08028
(609) 881-6677

Hudson
P.O. Box 6118
Hoboken, NJ 07030
(201) 782-6112

Hunterdon
84 Park Avenue
Flemington, NJ 08822
(201) 782-6112

Mercer
652 Whitehead Road
Trenton, NJ 08648
(609) 394-5000

Middlesex
2303 Woodbridge Avenue
P.O. Box 601
Edison, NJ 08818
(201) 985-9566

Monmouth
1540 Route 38
Suite 303
Wall Township, NJ 07719
(201) 280-2323

Morris
120 Washington Street
P.O. Box 2325
Morristown, NJ 07960
(201) 538-5336

Ocean
1540 Route 38
Suite 303
Wall Township, NJ 07719
(201) 370-0770

Passaic/Lakeland
843 East 27th Street
Paterson, NJ 07513
(201) 278-4184

Salem
19 N. Main Street
Woodstown, NJ 08098
(609) 769-0150

Somerset
120 Finderne Avenue
Bridgewater, NJ 08807
(201) 725-4664

Sussex
100 Main Street
Newton, NJ 07030
(201) 383-1334

Union
507 Westminster Avenue
P.O. Box 815
Elizabeth, NJ 07207-0815
(201) 354-7373

Warren
Route 46, RDI
P.O. Box 342
Oxford, NJ 07863
(201) 453-2103

American Heart Association Offices

National Office:

American Heart Association
7320 Greenville Avenue
Dallas, TX 75231
(800) 527-6941

New Jersey State Office:

American Heart Association of New Jersey
P.O. Box 1900
2550 U.S. Route 1
North Brunswick, NJ 08902
(201) 821-2610

New Jersey Regional Offices:

- In Hunterdon, Middlesex, Morris, Somerset, Sussex, and Warren Counties, call (201) 685-1118.
- In Mercer, Ocean, and Monmouth Counties, call (201) 222-2525.

- In Atlantic, Burlington, Camden, Cape May, Cumberland, Gloucester, and Salem Counties, call (609) 881-1860.
- In Bergen and Passaic Counties, call (201) 791-5800.
- In Essex, Hudson, and Union Counties, call (201) 376-3636.

American Lung Association Offices

National Office:

American Lung Association
1740 Broadway
New York, NY 10019-4374
(212) 315-8700

New Jersey State Office:

American Lung Association of New Jersey
1600 Route 22 East
Union, NJ 07083
(201) 687-9340

New Jersey Regional Offices:

Bergen, Essex, Morris, Passaic, Sussex, and Warren Counties
American Lung Association of New Jersey
Northern Regional Office
14-25 Plaza Road
Fair Lawn, NJ 07410
(201) 791-6600

Hudson, Monmouth, and Union Counties
American Lung Association of Central New Jersey
206 Westfield Avenue
Clark, NJ 07066-1539
(201) 388-4556

Burlington, Hunterdon, Mercer, Middlesex, and Somerset Counties
Delaware-Raritan Lung Association
P.O. Box 2006
Princeton, NJ 08543-2006
(609) 452-2112

Atlantic, Camden, Cape May, Cumberland, Gloucester, Ocean, and Salem Counties
American Lung Association of New Jersey
Southern Regional Office
10 West Main Street
Mays Landing, NJ 08330
(609) 625-0101

APPENDIX E

General Information

1. STANDARD U.S. FEDERAL REGIONS

Region I Connecticut, Maine, Massachusetts, New Hampshire, Rhode Island, Vermont

Region II New Jersey, New York, Puerto Rico, Virgin Islands

Region III Delaware, District of Columbia, Maryland, Pennsylvania, Virginia, West Virginia

Region IV Alabama, Florida, Georgia, Kentucky, Mississippi, North Carolina, South Carolina, Tennessee

Region V Illinois, Indiana, Michigan, Minnesota, Ohio, Wisconsin

Region VI Arkansas, Louisiana, New Mexico, Oklahoma, Texas

Region VII Iowa, Kansas, Missouri, Nebraska

Region VIII Colorado, Montana, North Dakota, South Dakota, Utah, Wyoming

Region IX Arizona, California, Hawaii, Nevada, American Samoa, Trust Territories of the Pacific, Guam, Northern Marianas

Region X Alaska, Idaho, Oregon, Washington

2. STEPS FOR RESPONDING TO A CHEMICAL EMERGENCY

Source: Chemical Manufacturers Association

General Guidelines

1. *Approach cautiously.* Resist the urge to rush in. You cannot help others until you know what you're getting into. Look for the cues of hazardous materials.

2. *Identify hazards.* Placards, container labels, shipping papers, and/or knowledgeable persons on the scene are your primary sources of information. Use them and consult the Department of Transportation's Emergency Response Guidebook.

3. *Secure site.* Without entering the hazard area, do what you can to isolate the site to assure the safety of persons and the environment.

4. *Obtain help.* Hazardous materials change the rules. Call for assistance from trained experts via Chemtrec at (800) 424-9300. In New Jersey, call the New Jersey Department of Environmental Protection's 24-hour Environmental Protection Hotline at (609) 292-7172 to notify responsible agencies.

5. *Carefully consider site entry.* Efforts to rescue persons or protect property or environment must be carefully weighed against the possibility that you will become part of the problem. If you must enter the site, use appropriate protective equipment.

Suspected Radioactive Material

Source: New Jersey Department of Environmental Protection

1. Do not touch the container, nor the immediate area. Wear protective clothing and equipment.

2. Use your SURVEY METER to determine if radioactive material is present. Use the most sensitive scale.

3. Restrict access to the area where the material is located. Use rope, signs, fences, etc.

4. If you measure any reading on your instrument that is twice background, then you can assume that radioactive material is present. (Background can be taken any place not in the vicinity of the suspected radioactive material.)

5. Look for the following information on the container and record:
   - name of radioactive material (Tc99, Cs137, I131, Co60, etc.)
   - type of material (solid, gas cylinder, liquid vial)
   - quantity of activity of material (in μCI [microcuries] or CI [curies])
   - date of calibration

6. In New Jersey, notify the New Jersey Department of Environmental Protection by calling the 24-hour Environmental Protection Hotline at (609) 292-7172. The Department will then contact appropriate authorities, including State Police and local authorities.

   Give your name, organization, telephone number, location, and any pertinent information from Items 1 through 5.

New Jersey Procedures

If petroleum or other hazardous substances are leaked or discharged from your facility, you must call the DEP immediately. The National Response Center (800) 424-8802 must be notified if the discharge exceeds the reportable quantity for that chemical.

The owner, operator, or person in charge should phone the DEP Environmental Protection Hot Line at (609) 292-7172. When you call you should report:

1. the type of substance

2. the estimated amount discharged, if known

3. the location

4. the actions you are taking to contain, clean up, or remove the substance, if any

DEP, in turn, will contact their Emergency Response Unit, the State Police, and local emergency personnel.

The owner or operator must, within 10 days, make an inspection and file a *written* report (containing the four items listed above) with the local municipal government and the local board of health. Failure to do this, or lying on the report, can lead to a fine of up to $25,000 per day per violation. Written follow-up to DEP should be addressed to:

Department of Environmental Protection
Division of Environmental Quality
Bureau of Communications and Support Services
CN 411
Trenton, NJ 08625

For any type of chemical emergency information, the Chemical Manufacturers Association has a 24-hour emergency hotline:

Chemtrec—(800) 424-9300

For more details on New Jersey's environmental laws, order "The Role of Police in Environmental Protection" by Marshall Staley, Theodore B. Shelton, and Barie E. Kline (Cooperative Extension Service, Cook College, Rutgers, The State University of New Jersey), Bulletin #420, $1. (201) 932-9762

For information on emergency response coordination, contact:

Bureau of Emergency Response
New Jersey Department of Environmental Protection
CN 027
Trenton, NJ 08625
(609) 633-2658

A valuable resource for chemical emergencies is the *Emergency Response Guidebook,* by the U.S. Department of Transportation. (See "Health and Safety Core Reference Library," Chapter 5, Section 1.)

3. GLOSSARY OF OCCUPATIONAL HEALTH TERMS

ACGIH   American Council of Government and Industrial Hygienists. A nongovernmental group that develops Threshold Limit Values (*see* TLVs).

acute   having a sudden onset, sharp rise, and short course.

carcinogen   a substance or agent producing or inciting cancer.

chronic   of long duration. Usually refers to a disease that progresses slowly and continues for a long time.

congenital   existing at, or dating from, birth—acquired during development in the uterus.

Consumer Product Safety Act   provides the Consumer Product Safety Commission with power to regulate hazardous substances related to consumer products, including the setting of standards for performance, composition, design, construction, finish, or packaging.

decibels (dB)   unit for expressing the relative intensity of sound, on a scale from zero for the average least perceptible sound to about 130 for the average pain level.

Delaney Amendment   clause added to the Federal Food, Drug, and Cosmetic Act in 1958 which states that any substance that causes cancer in animals must not be added, in any amounts, to food.

dose-response how a biological organism's response to toxic substance changes as its exposure to the substance changes. For example, a small dose of carbon monoxide will cause drowsiness; a larger dose can be fatal.

epidemiology the study of the prevalence of human diseases in different places and the circumstances that accompany high rates of disease.

ergonomics study of human factors in the design and operations of machines and the physical environment, most often in the work setting.

exposure uptake into the body of a hazardous material. The most common routes of exposure are dermal (skin), oral (mouth), and inhalation (breathing).

hazardous substances subsets of the all-encompassing term, "hazardous materials," they are substances or mixtures that are toxic, corrosive, flammable, explosive, or reactive. In general, they pose a risk to living things and/or the environment.

incidence the rate of occurrence—for example, the number of new cases of a disease occurring in a particular period of time.

industrial hygiene the recognition of environmental factors associated with work and work operations, the evaluation of how these factors affect workers' health and well-being, and the prescription of the methods to control or eliminate such factors or minimize their effects.

mutagen a substance that tends to increase the frequency or extent of changes in hereditary material, involving a physical change in chromosomes or biochemical change in genes.

NIOSH National Institute for Occupational Safety and Health. NIOSH was established to provide research to identify, evaluate, and control work-related illness and injury; to disseminate information to workers, employers, and health professionals; and to train occupational safety and health professionals.

Occupational Safety and Health Administration (OSHA) Created in 1970 to encourage reduction in job hazards, OSHA maintains and enforces standards of health and safety; establishes rights and responsibilities for employees and employers; maintains records on job-related injuries and illnesses; and conducts research.

organic chemicals chemical compounds containing carbon. Historically, organic compounds were those obtained from vegetables or animal sources. Today, many organic chemicals are synthesized in the laboratory.

personal protective equipment (PPE) equipment or clothing intended to protect workers from worksite hazards that cannot be eliminated. May include eye, face, or head protection; hearing protection; respirators; lifelines and safety belts; foot protection; and special work clothing, such as gloves or special suits.

ppb parts per billion.

ppm parts per million.

prevalence   the percentage of a population that is affected with a particular disease at a given time.

Right to Know   law that applies to work settings (that use hazardous chemicals), requiring certain activities to keep workers informed of potential health risks and their control. Also, "Community Right To Know" is designed to provide similar information to the community at large.

sampling (air, water)   small amounts of air or water are obtained and analyzed to determine what levels of different substances are contained in the air or water at a certain location. This is often done so that knowledge of potentially hazardous exposures can be obtained and control measures can be taken.

solvent   a substance (usually a liquid) capable of dissolving one or more other substances.

spirometry   measurement of air entering or leaving the lungs, in order to test lung function.

synergistic effect   the simultaneous action of separate agents which, together, have a greater total effect than the sum of their individual effects.

teratogen   a substance or agent that could change an embryo as it is developing during pregnancy, causing a birth defect. Teratogenic effects are not passed on to future generations.

Threshold Limit Values (TLVs)   levels of airborne contaminants that most people can be exposed to without adverse effects. TLVs are periodically revised and updated by ACGIH, which also establishes them, and are published annually.

toxic   poisonous; or of, relating to, or caused by a poison.

toxicology   the science of the nature and effects of poisons, their detection, and the treatment of their effects.

Toxic Substance Control Act (TSCA)   signed into law in 1976, TSCA is intended to "ensure that chemical substances and mixtures are regulated in a manner that ensures that they do not present an unreasonable risk of injury to health or the environment." It is administered by the EPA.

Index

acute effects of hazardous substances 66
acute exposure to hazardous substances 66
air monitoring 65
air-purifying respirators 82
air samples 5, 68
air-supplying respirators 84
American Board of Preventive Medicine
    61
American Cancer Society 159, 189–192
American Conference of Governmental
    Industrial Hygienists 131
American Heart Association 189, 192–193
American Industrial Hygiene Association
    122, 132
American Lung Association 159, 181,
    189, 193
American National Standards Institute 132
American Society of Safety Engineers 180
approved respirators 85
Association of Occupational and
    Environmental Clinics 123

biological contaminants 4
biological monitoring 62
bulk samples 68
Business Library of Newark Public Library
    129–130

cancer 2
Center for Occupational Hazards 181
CERCLA *See* Comprehensive
    Environmental Response,
    Compensation and Liability Act
chemical emergencies 195
chemical use 4
chronic effects of hazardous substances 66
chronic exposure to hazardous substances
    66
Clean Air Act (CAA) 45, 165
Clean Water Act (CWA) 42, 165

Section III (Oil Spills) 51
Committees on Occupational Safety and
    Health 134
Comprehensive Environmental Response,
    Compensation and Liability Act 53–
    55, 168
COSH *See* Committees on Occupational
    Safety and Health

DEP *See* New Jersey Department of
    Environmental Protection
disease monitoring tests 62
DOH *See* New Jersey State Department of
    Health
dose 65
dusts 64

ear protection 79–80
ECRA *See* Environmental Cleanup
    Responsibility Act
Emergency Planning and Community
    Right-to-Know Act *See* SARA Title
    III
employee responsibilities 13
employee rights 12, 33, 63, 68
employer responsibilities 13, 29
enclosure 6
Environmental Cleanup Responsibility Act
    59–60, 170
Environmental Hazards Management
    Institute 181–182
Environmental Protection Agency
    RCRA/Superfund Hotline 167
    regional offices 171
    Region II directory 172
    SARA Title III Hotline 157, 169
    Small Business Hotline 164
    Toxic Substances Control Assistance
    Office 165

EPA *See* Environmental Protection Agency
exposure limits 70
exposure record 63
eye protection 75–76

face protection 75–76
foot protection 78
fumes 64

gases 65

hazard minimization 5
Hazardous Substance Fact Sheets 61, 67, 152
head protection 74–75
health care costs 95–96
health care providers 61
housekeeping 6

identifying a hazard 66
illness and injury costs 98–102
illness cost trends 98–100
industrial hygiene monitoring 68
ingestion 65
inhalation 65
Institute of Management and Labor Relations Library 130
International Brotherhood of Painters and Allied Trades 182
International Molders and Allied Workers Union 182
isolation 6

John Cotton Dana Library 130

labels 67
laboratories and analytical methods 70
laboratory accreditation 122
latency 66
lifelines 78
liquids 64
local effects of hazardous substances 66

Material Safety Data Sheets 19, 61, 67, 155, 157–158
mechanizing a process 7
medical records 63

mists 64
monitoring
    environmental monitoring 5
    medical surveillance 5
MSDSs *See* Material Safety Data Sheets

National Cancer Institute 189
National Institute for Occupational Safety and Health 9, 132
    Educational Resource Centers 178–180
    Health Hazard Evaluation Program 121–122
    regional offices 149
National Pesticide Telecommunications Network 164
National Safety Council 133, 180–181
New Jersey Department of Environmental Protection 170, 173
New Jersey Office of Small Business Assistance 128–129
New Jersey Public Employees Occupational Safety and Health Act 63, 136
New Jersey Right to Know Act 152
    background 21
    education and training 31
    employers covered 21
    enforcement and penalties 33
    funding 29
    health records 31
    labeling 32
    making information available to employees 31
    research and development laboratories 22
    SIC Codes covered 23
    surveys 29
    trade secrets 33
New Jersey Small Business Development Center 127–128
New Jersey State Board of Medical Examiners 63
New Jersey State Department of Health 136
NIOSH *See* National Institute for Occupational Safety and Health

Occupational Safety and Health Act 9
Occupational Safety and Health
    Administration 9
  appeals 15
  area offices 142
  citations/penalties 15
  closing conference of inspection 12
  Consultation Program 108–112
  general duty clause 14
  inspection exemption 111
  inspection requests 11
  Notice of Contest 15
  Notification of Failure to Correct
    Alleged Violations 15
  opening conference of inspection 11
  OSHA 200 Log of Injuries and Illness 13
  regional offices 141
  schedule of inspections 10
  standards 10, 15
  Supplementary Record of Injury or
    Illness 14
  violations 14
  Voluntary Protection Program 117–120
  walkaround inspection 12
Occupational Safety and Health Review
  Commission 9
Office of Disease Prevention and Health
  Promotion 160, 188
OSHA See Occupational Safety and Health
  Administration
OSHA Hazard Communication Standard 2,
  16
  chemicals covered 18
  employee training 20
  enforcement 20
  history 17
  industries covered 17
  labeling 18
  trade secrets 20
  written program 18

permissible exposure limits 70
personal protective equipment 7
physical environment 4
protective clothing 76–78
Public Employees Occupational Safety and
  Health Act See New Jersey Public

Employees Occupational Safety and
  Health Act

qualitative fit tests for respirators 85
quantitative fit tests for respirators 86

radioactive material 196
RCRA See Resource Conservation and
  Recovery Act
Resource Conservation and Recovery Act
  48–51, 167

safety belts 78
safety inspection of the workplace 3
safety policies 7
sampling devices 69
sampling preparation 69
SARA Title III 34, 56–57, 168
  hazardous substances inventory
    reporting 35
  Local Emergency Planning Committee
    35
  State Emergency Response
    Commissions 34
  toxic release inventory reporting 35
SCCA See Spill Compensation and Control
  Act
Service Corps of Retired Executives
  (SCORE) 125–127
short-term exposure limit 71
skin contact 65
Small Business Administration 125–126
smoking in the workplace
  cessation programs 159–162
  health effects 36
  New Jersey legislation 37
  policies 159–162
solids 64
Spill Compensation and Control Act 58–
  59, 170
stress 5
substitution 6
Superfund See Comprehensive
  Environmental Response,
  Compensation and Liability Act
Superfund Amendments and
  Reauthorization Act See SARA Title
  III

systemic effects of hazardous substances
66

Threshold Limit Values 70
   ceiling TLV 71
time-weighted averages 71
Toxic Catastrophe Prevention Act (TCPA)
   57–58, 169
Toxic Substances Control Act (TSCA) 47,
   165
toxic substances tests 62

union involvement in safety and health 92–
   93

vapors 64
ventilation 6

wipe samples 69
Women's Occupational Health Resource
   Center 181
work environment 3–7
Worker's Institute for Safety and Health
   182
worksite wellness programs 96–98